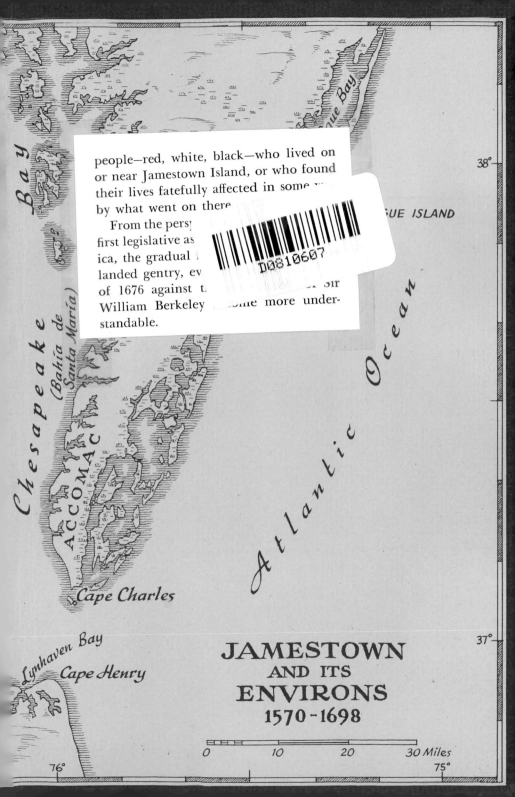

people—red, white, black—who lived on
or near Jamestown Island, or who found
their lives fatefully affected in some [...]
by what went on ther[...]

From the pers[...]
first legislative as[...]
ica, the gradual [...]
landed gentry, ev[...]
of 1676 against t[...] Sir
William Berkeley [...]ome more under-
standable.

JAMESTOWN
AND ITS
ENVIRONS
1570-1698

Jamestown, 1544-1699

Jamestown
1544-1699

Carl Bridenbaugh

New York   Oxford
OXFORD UNIVERSITY PRESS
1980

Copyright © 1980 by Carl Bridenbaugh

Second printing, 1980

Library of Congress Cataloging in Publication Data

Bridenbaugh, Carl.
Jamestown, 1544-1699.

Bibliography: p.   Includes index.
1.   Jamestown, Va.–History.   I.   Title
F234.J3B7   975.5'4251   79-13989
ISBN 0-19-502650-0

Printed in the United States of America

To the Memory of My Friend
the Late Leonard Ware
Journalist, Naval Officer,
Indian Official

Preface

Here, reader, is no chronicle of saints and patriots nor yet is it a history of white and "tawny" devils. It is a record of both, a tale of arrogance and fear, of suffering, violence, and death that suffused a spot called Jamestown in Virginia nearly four centuries ago where the American people had its beginnings.

We lack detailed records that would enable the historian to recount the story accurately and impartially of both the native Indians and the Englishmen from overseas. Much of what has been handed down or written has been overlaid with romanticism, local pride, or racial prejudice, which have blurred or distorted our history. Good history is never a tale of living happily ever after, and the American success story was seldom the lot of more than a few of the colonists of early Virginia. An overwhelming number succumbed, and the remainder merely survived. In this story of Jamestown and the Virginians—red, white, and black—I have tried to relate the best and the worst in the knowledge that both entered into the experience of the first Americans and have become, to a certain extent, a part of the national heritage.

This work is addressed to the general reader who wishes

to learn something about the history and significance of the place. For those persons who may want to check the reliability of the text I have, immediately following it, cited all of the authorities for quotations, facts, and figures for my statements and conclusions. To assist my audience further in placing events, I have provided a chronology of Jamestown in Appendix 1.

The map on the endpapers and the illustrations are an integral part of the book. Beneath each illustration is a short caption, and in the list of illustrations this is repeated and followed by a longer description of the picture and its source, for these contemporary representations should be read with the text of the book. They are intended to inform as well as to please the reader.

My special thanks are due to Father Albert J. Loomie, S. J., of Fordham University for supplying his own translation of a very important document from the Spanish Archives, and to him and his colleague Father Clifford M. Lewis, S. J., whose book on the Spanish mission in Virginia I have drawn upon so heavily. I have not always followed their interpretation of the documents, but I do commend their industry and scholarship.

For assistance rendered during the preparation of this work, I wish to thank Heath Pemberton of the National Park Service. My friend John Melville Jennings of Toano encouraged me and supplied essential materials. Parke Rouse, Jr., kindly granted permission to include materials from the publications of the Jamestown/Yorktown Foundation. At Providence, Samuel Hough and the staff of the John Carter Brown Library opened their unrivalled collections and showed me every courtesy. I am always in debt to Jeannette Black for understanding and advice. David Beers Quinn first directed my attention to Opechancanough, though he should not bear any responsibility for my interpretations. At a most critical time Alison Quinn encouraged me. As always, Roberta Bridenbaugh's assistance and criticism have made this a better book. Our residence for five years at Williamsburg deepened our understanding of early Virginia and nearby Jamestown Island, and allowed us to develop a fondness and reverence for that solitary site.

Providence　　　　　　　　　　　　　　　　　　　　c. b.
April 3, 1979

Contents

Part I
Momentous and Tragic Beginnings

I   The English Invasion   3

II   Opechancanough, the Powhatans,
and the Europeans, 1561-1644   10

III   John Rolfe, Pocahontas,
and "that chopping herbe of Hell,"
Tobacco   34

IV   Life and Death at Jamestown   44

V   God and Man at Jamestown   61

VI   Self-government and Self-interest
in a Planters' Parliament   76

VII   Bacon's Uprising,
Lawrence's Rebellion   89

Part II
"James Cittie in Virginia"

VIII   The Jamestown Community:
A Conspicuous Failure   107

IX   The People, the Site, and the Port   118

X   Jamestown: Symbol for Americans   150

Notes   155

Appendices:
  I   Jamestown Chronology, 1544-1699   177
  II   A Note on Archaeology and Restoration
       at Jamestown   185
  III For Further Reading   187
  IV   Governors of Virginia to 1699   190

Index   193

List of Illustrations

NOTE: The four illustrations listed below were published in England or Germany between 1624 and 1675. The work of European artists and engravers, they reflect the Virginia scene as seen and conceived by white men and are, therefore, "biased sources." Readers should keep this in mind, but, at the same time, understand that they are nevertheless valuable records of actual events —the only ones we possess. Perhaps their primary worth is in showing that the white men from overseas always viewed the native Americans through European spectacles.

Endpapers   Jamestown and Its Environs, 1570-1698

Plate 1.   Don Luis de Velasco Killing the Jesuits, 1571   p. 18
Drawn by C. Screta and engraved by Melchior Küssell and used to illustrate Father Mathias Tanner's *Societas Iesu usque ad Sanguina et Profusionem Vitae Militans* (Prague, 1675), p. 450. Don Luis (Opechancanough) is portrayed as the huge Indian of the literary accounts, though the artist gave him a cleaver-type axe and a costume known only in Europe at that time. The inscription reads. "F. Juan Bapt: de Segura, Gabriel Gómez, Pedro Linarez, Sanctus Sauellius, Christoph. Rotundo, Spanish Jesuits, were murdered in Florida [Virginia] for the Faith of Christ, in the year 1571 on the 8th of February." Father Segura, with rosary and crucifix in hand and head bowed in prayer, displays

the traditional and conventional pose of the Catholic martyr, whose thoughts are focused upon eternity, not the murderer. (Courtesy of the John Carter Brown Library)

Plate 2.   Captain John Smith Threatens Opechancanough, 1608
p. 21
Again, the chieftain's great stature is stressed. The fighting shown in the background did not take place on this occasion. This vignette is one of eight designed and engraved by Robert Vaughan of London for Smith's map of "Ould Virginia" in the *Generall Historie of Virginia* (London, 1624), facing p. 20. Captain Smith worked with Vaughan, who thereby was more accurate in his representations than most of the Europeans. (Courtesy of the John Carter Brown Library)

Plate 3.   The Massacre of 1622   p. 30
This scene, crowded with humanity as in a Breughel painting, tells little of the event, save that it was a tragedy for the English. Once more, the Indians are portrayed as larger than their victims, as they wield daggers or poniards rather than the native stone hatchets. We know, too, that not a few of the Indians had firearms. The buildings shown are such as the designer and publisher knew at Frankfurt am Main. Included by the celebrated Theodor de Bry in *America Pars Decima*, Part XIII (1634), facing p. 28, this depiction of the Massacre is significant for the image it fixed in the European mind. (Courtesy of the John Carter Brown Library)

Plate 4.   Pocahontas in 1616   p. 39
During her visit to London at the age of twenty-one, Pocahontas sat for her portrait. It was engraved and sold widely by Simon de Passe, but the artist is unknown. She is shown looking sober in her English finery, revealing nothing of the playful youngster who had turned cartwheels in Jamestown years before. John Smith printed de Passe's engraving in the *Generall Historie of Virginia* (London, 1624), facing p. 112. (Courtesy of the John Carter Brown Library)

Part I

Momentous and Tragic Beginnings

I

The English Invasion

"There it comes!"

From a perch high in a tree on the river's bank, an alert Indian lookout shouted down to his companions crouching in a clump of bushes. Excitedly these watchers talked among themselves in an Algonkian dialect as they sought to determine who and what it was they had discovered moving up the Powhatan River (soon to be called the James) on May 4, 1607. Wowinchopunk, the "King of the Paspaheghs,"* having been warned of the arrival of three great ships in Chesapeake Bay, had stationed small parties at distances within easy communication of each other along the north shore of the river all the way upstream, from Kecoughtan at its mouth to the King's principal village lying beyond its confluence with the Chickahominy.

* *Paspahegh* was the Algonkian name for Jamestown Island when there was an Indian settlement there. It was also the name of an Indian tribe in the neighborhood of the island when the English came; and near the confluence of the Chickahominy and James rivers was the tribe's principal village of Paspahegh. Paspahegh was a peninsula in 1607, but in later years it became an island. To avoid confusion, in this work the spot will be called "Jamestown Island," the modern usage.

What the lookouts had spotted was a small open vessel—known as a shallop—aboard which, as it drew nearer, they could make out several clothed and bearded fair-skinned men. The craft sailed upriver past an island called Paspahegh (it would shortly be renamed Jamestown); eight days later the Indians spied the shallop coming downstream with the tide. The following day three large ships ascended the river as far as the island and then moored inshore with lines made fast to trees on the bank; the next morning more than a hundred white men went ashore. After unloading quantities of supplies and other goods, most of them set about making a fortification while others seemed "to watch and ward as it was convenient."[1]

Why have they come? What are they looking for? Are they going to stay? These and many other questions must have been discussed by the startled Paspaheghs. And well might they have been suspicious and fearful that these strange men were planning to settle in their midst. Although neither the white men nor the red realized it at the time, this invasion of North America signaled the inception of a profound and lasting human tragedy.

The white men from the three ships were Englishmen, whose reasons for coming were many and complex. From the epoch-making voyage of discovery made by Christopher Columbus in 1492 to the end of the sixteenth century, Spanish kings ruled exclusively and, in general, effectively over the peoples and lands in the New World called America north of the equator as far as the present state of Georgia and in the great islands of the Caribbean. If the monarchs of Western Europe strove for dynastic eminence, their subjects identified themselves by their intense religious convictions: the abortive conquest of the British Isles attempted by the Spanish Armada and the long sea-war from 1588 to 1604 was as much a conflict of Roman Catholic Spaniards against English Protestants as a contest for national and maritime supremacy. Likewise the revolt of the Low Countries against King Philip II of Spain in 1568 was a struggle of the Protestant burghers of Holland and Zeeland against His Catholic Majesty quite as much as it was a Dutch bid for independence and commercial advantage.

The King of Spain not only dominated the lands to the west but he controlled the shipping across the Atlantic. Quantities of gold, silver, and exotic products, plundered from the helpless natives of the New World and transported back to the port of Seville, poured into the coffers of the royal house. Such a rich treasure could not but excite the envy and greed of the French, English, and his own rebellious subjects in the Netherlands. All three of these peoples had only recently embarked upon maritime enterprise on a large scale, and inevitably mariners from these countries became interlopers, challenging the Spanish monopoly "beyond the line." Though thinly veiled, their voyages were those of pirates. During wartime, they sailed westward as privateers to wage a semi-official *guerre de course*, as the French called it.[2]

In addition to the compelling desire to share in the wealth of America at the expense of Spain, the English hoped to discover an all-water route to China, such as the Portuguese and Spanish already enjoyed. Even before Columbus discovered America, merchants at Bristol had dispatched ships westward (1480-94) to begin the long, fruitless search for a Northwest Passage. After 1580 English trade expanded rapidly, together with a marked increase in shipping. During the years of war with Spain, English privateers, financed by merchants at London, harried Spanish shipping in the West Indies with the blessing of Good Queen Bess and her ministers. Not only did many of these ventures prove profitable, but the captains and their crews grew familiar with the transatlantic routes of commerce and the waters of the Caribbean; they also noted the progressive weakening of the Spanish government and its defenses in the New World.

When King James I made peace with Spain in August 1604, the merchant backers and the privateersmen were ready and eager to return to peacetime oceanic trade if the Spanish would recognize their right to traffic with His Catholic Majesty's American colonies. Although they failed to win that right in the treaty of peace, they nevertheless, illegally, continued trading in the Caribbean. "Christopher Newport of Limehouse, Mariner," had made his first voyage to the West Indies as a privateersman in 1589 and had subsequently participated in other ventures, partly for trading, partly to take prizes. At the age of forty-six, widely known as "a Mariner well practised for the Westerne parts of America," he was given command of the first English Virginia expedition in 1607.[3]

To the openly expressed intentions to increase shipping and expand trade, to acquire the newly available liquid wealth (instead of landed estates), to participate in the discovery of unknown lands, and to traffic overseas—all overlaid by national and religious impulses that fired their hatred of Spain—must be added the stimulus of a genuine patriotic spirit among Englishmen of all ranks. That which drove or beckoned them overseas was not so much the desire for money—very little was made for a long time—as it was the belief that the future greatness of their country depended upon such undertakings. "It was a time of more loss than profit," a great historian has written, "of more misery than glory . . . The way of these men was hard, and their reward small . . . They themselves closed their eyes on failure."[4]

The two attempts of Sir Walter Raleigh to establish an English settlement in Virginia at Roanoke Island (now part of North Carolina) in 1585 and 1587 had failed for a variety of reasons, but one thing was proved: successful colonizing required a much larger investment than any one person could command. After several probing voyages had been made, including one by Captain Newport in 1605, several London merchants—some of whom had been deeply involved in privateering—a number of soldiers who had fought in the Netherlands, and some country gentlemen from the West of England decided to form a trading company (much like the recently organized East India Company) to establish colonies in North America. The list of investors in the Company indicates that this was a national enterprise. On April 10, 1606, King James granted them a charter as The Virginia Company of London,* which authorized the sending out of "the *first colony*" to the coast of Virginia.[5]

The idea of a permanent English colony in the New World thus passed from dream to reality. The concept had been popularized throughout the land by what James A. Froude felicitously called the prose epic of the modern English nation. At London in 1598, the first of three great folio volumes came off the press, and in 1600 appeared *The Third and Last Volume of the Voyages, Navigations, Traffiques and Discoveries of the English Nation, and in some few places, where they have not been, of strangers, performed within*

---

* The correct name for this body was The Virginia Company of London. In these pages it will also be referred to as the *Virginia Company*, the *London Company*, or merely as the *Company*.

and before the time of these hundred yeeres, to all parts of the Newfound World of America . . . Collected by Richard Hakluyt Preacher. Readers of Hakluyt's *Voyages* satisfied their curiosity aroused by the talk they had heard in public houses of new discoveries and travels to distant places by their countrymen, and at the same time they found convincing the editor's sound arguments of the national need for colonizing in America. Proof of the Anglican minister's importance to this movement was his membership among the applicants for a charter for the Virginia Company. More than any other person he had prepared the public mind for action on Western planting.

Well aware that a permanent settlement, not just a trading post, was necessary for their purposes, the officers of the Virginia Company laid their plans accordingly. Emigrants were available in substantial numbers, not merely in London but almost anywhere in Britain, for the general prosperity, the result of England's surge ahead among the nations of Western Europe, was not widely distributed. Only a very small minority enjoyed the good life. Actually there were thousands of vexed and troubled Englishmen from whose ranks settlers could be recruited.[6]

Contrary to what we may have read in history books, the Company was not, in 1606, proposing to plant a new nation. The officers' objective, while a novel and large-scale experiment, was much more modest and prosaic: they intended to explore the scene and build a fort as a preliminary to a larger settlement. Most of the first colonists sent out would be expected to fish and grow crops to feed themselves, and to trade with the Indians. Others were to survey carefully the Chesapeake Bay and tributary rivers with an eye to discovering a passage to Cathay, and to trace out and map the still unknown seacoast from the Savannah River to the Bay of Fundy. In the course of these activities, one subsidiary objective was to find the survivors of the "Lost Colony of Roanoke" about whose existence many rumors still circulated; another, much discussed—and sincerely so—was the proper treatment and eventual Englishing and Christianizing of the Indians of Virginia. When Lord Ellesmere was asked in 1609 why there had been "such a bustle . . . about sending [people] to Virginia," he was said to have replied, "at first we always thought of sending people little by little, but now we see what we should do is establish ourselves all at once" on a large scale to warn off the Spaniards and the natives.[7]

It was in conformance with these plans that the advance contingent of the Great Migration of the English in the seventeenth century sailed for Virginia under the command of the "well practised" Captain Christopher Newport. After guiding it unerringly by way of the Azores and entering the great Bay of Chesapeake without incident, he began at once to examine the valley of the James. It was his ships—small by European standards—*Susan Constant* (100 tons), *Godspeed* (40 tons), and *Discovery* (a pinnace of 20 tons), from which the amazed Paspaheghs saw a hundred and more colonists disembarking on May 14, 1607.

It was a well-planned and well-carried-out expedition. A fort was immediately constructed to guard the outpost against a Spanish approach by sea and to keep it safe from aboriginal attack from the land. To the Algonkian tribes the arrival of the English and evidence of their determination to stay in Virginia were profoundly shocking. For neither side was there the least trace of the romantic in this portentous event; for all of them, red and white, it was a grim and serious business.[8]

II

Opechancanough, the Powhatans, and the White Europeans, 1561-1644

Before the settlement of Jamestown on the island in the James River in 1607, there had once been an Indian village, Paspahegh, on the very same spot. The Spanish had also landed near there, and for ages past numerous Algonkian-speaking tribes of Indians had lived thereabout. The men from Britain came with the avowed intent of propagating the Christian religion and teaching the "savages" human civility. They harbored not the slightest suspicion that the natives had long since encountered and dealt with white men and that they both mistrusted and feared these Christians who crossed the seas in great ships. And what was more important, from the very beginning these uninvited strangers never understood the extent of the hazard involved in the confrontation of the two races.[1]

From 1607 to our own time, Powhatan, the despotic ruler over many tribes in the neighborhood of Jamestown Island, has been looked upon as the foremost native leader in seventeenth-century Virginia. Today, however, we may properly inquire whether his elder brother, *Opechancanough*, did not greatly surpass him in talents and capacity for leadership and should also rank very high in any list of the most famous Indians of American history. The remarkable story of the life and adventures of this early American Indian patriot—untold until now—epitomizes the first century of contact between the natives and the English, which had its beginnings at Jamestown.*

This brother, O-pe-chán-can-ough, was believed to be over a hundred years old in 1644; he must, therefore, have been born about 1544. Early in the spring of 1561, when Opechancanough would have been a tall, attractive young man of sixteen or seventeen, Pedro Menéndez de Avilés, in command of two Spanish ships returning to Spain from Havana, discovered and entered Bahía de Santa María (Chesapeake Bay). No sooner did some alert Indians see the strangers lower their sails and anchor than they paddled out in their canoes and boarded the flagship. "Among these Indians came a chief who brought his son . . . who was of fine presence and bearing." The "Admiral" generously gave the natives food and clothing. Being particularly attracted by the youth, he asked the father's permission to take him across the Atlantic so "that the King of Spain, his lord, might see him." In return, Menéndez "gave his pledged word to return him with much wealth and many garments." The request being granted,

---

* The reader must understand that the story of Don Luis/Opechan-canough related below cannot be proved. It is, however, a reasonable, workable, and plausible hypothesis into which the known facts fit nicely, an interpretation that explains nearly all of the matters concerning Opechancanough and his brother, the Powhatan, that have hitherto been obscure, preventing our understanding of many events in the early history of Virginia. In the present chapter references are given only for quotations and facts in the text. The argument has been developed in a much longer, more detailed, and fully documented study of the subject in *Early Americans* to be published by Oxford University Press in 1981.

the young Indian embarked upon the kind of experiences that the Spaniards of the day referred to as *picaresco*.[2]

The ship reached Cadiz in the late spring or early summer of 1561. Pedro Menéndez de Avilés promptly introduced the American at the court of King Philip II as "a *cacique* or important lord from . . . Florida." His imposing stature, fine physique, and high intelligence astonished His Catholic Majesty, and his grandees were "very pleased with him." The monarch ordered an allowance for elegant clothes suitable to the noble condition of the chieftain and approved placing him with the Dominican friars at Seville for instruction in the Spanish language and the rudiments of Christianity.[3]

Both the King and Menéndez fully intended that, as promised by the nobleman, the Indian should ultimately be returned to his native land, but it was five long years before this was actually attempted. Meanwhile the neophyte made remarkable progress with Castilian and his sacred studies. All observers testified to his powerful intellect and also to the fact that he was "wily" and "crafty." In after years it became abundantly clear that during his stay in Spain he also accumulated, through observation and inquiry, a formidable body of information about the great number of Spaniards, their social life and ways, military strength, and technology. And, what was more significant for the future, he appears to have reflected deeply about the meaning and importance of much of his new knowledge. First from the Dominican fathers, later from the Jesuits, he learned that often what could not be won by force might be achieved by diplomacy, as well as the value of patience and long-range planning for great undertakings—both attributes signally wanting in American Indians dwelling north of the Rio Grande. Under the tutelage of the Dominicans, the still unnamed youth became a Christian.[4]

When Menéndez crossed to Mexico in command of the annual fleet in 1563, he took the Indian with him. The King had directed that the youth, who ardently desired to go home, be restored to his people, but when the "Admiral" attempted to carry out this order, and fulfill his pledge of 1561, the Archbishop of Mexico, fearing that the convert might lapse into his former devil worship, refused to give permission. Completely frustrated, Menéndez left the Indian there with the Dominicans, with whom he stayed for the next three years. During this time he was the protégé of Governor Don Luis de Velasco, who had acted as his godfather and bestowed his own name upon him. "Thus," observed the Jesuit Juan Rogel with typical European disdain, "the Indian son of a petty chief of Florida was called Don Luis."[5]

King Philip appointed Menéndez *adelantado* (conqueror) of Florida on March 20, 1565; two days later he issued a royal order to the authorities of New Spain to remand Don Luis to the custody of Menéndez "upon demand." From Havana on December 25, Menéndez dispatched the order to Mexico, where both the Archbishop and Governor reluctantly complied with it by sending Don Luis to San Mateo in Florida in the company of two Dominican friars.[6]

At San Mateo on August 1, 1566, the *Adelantado* ordered Fray Pablo de San Pedro and the other friar to proceed to Bahía de Santa María on a colonizing expedition. In one of his instructions he pointedly stated "that it is in the service of God, our Lord and of his Majesty that I send Don Luis, the Indian, to his country, which according to him is between the 36th and 39th degrees [north latitude] along the shore, and all the people of that territory are his friends and the vassals of his three brothers." Captain Pedro de Coronas and thirty soldiers joined the two Dominicans and sailed from Santa Elena (Parris Island) for Ajacán, the land bordering Bahía de Santa María. Bartolomé Barrientos wrote the next year that with them "went a brother of the cacique of that area. This man the *Adelantado* had taken from there six years before. He was very cultivated, showed good understanding, and was a good Christian. His name was Don Luis de Velasco, and he was sent to help the conversion of the natives." By this action, Pedro Menéndez de Avilés expected to redeem the pledge made to the father of Don Luis.[7]

La Trinidad, though guided by Don Luis, missed the entrance to Chesapeake Bay and anchored at 37° 30′, possibly in Chincoteague Bay. There a storm arose that blew the vessel as far south as 36°. After coasting along the Carolina Outer Banks and making a brief landing, the Portuguese pilot, Domingo Fernández, retraced his course northward as far as 36° 30′, but again a gale, four days in duration, drove *La Trinidad* far out to sea. Then, without consulting their Indian guide, the pilot, Captain Coronas, and the two friars decided to sail to Spain.[8]

On October 23, 1566, the ship made the harbor of Cadiz, and Don Luis went ashore to dwell once again in the Iberian Peninsula. The Dominicans, for reasons no longer known, dropped him off on November 20 in Seville on their way back to the court at Valladolid. Possibly Philip II directed that he be placed with the Jesuits for further education now that their order had taken the lead in missionary work. About 1568, according to Father Gerónimo de Oré, a Franciscan historian, "the *cacique* Don Luis was living in the [Jesuits'] house at Seville, advancing in the Spanish language, both in reading and writing, together with other branches of knowledge which they taught him." Sometime during this second Spanish sojourn the Indian "was made ready and they gave him the holy sacraments of the altar and Confirmation."[9]

Meanwhile, grieving over the failure to reach his home when he had been so near and desiring more than ever to return to Ajacán, the American, now twenty-three years old, approached his Jesuit mentors in the autumn of 1567 with a plan long-cherished, for "converting his parents, relatives, and countrymen to the faith of Jesus Christ and baptizing them and making them Christians as he was." Hearing that a party of Jesuits was about to depart for Florida in the hope of winning the natives to the Faith, Don Luis proposed to his rector that he send priests to his country and asked to accompany and assist them in the work of conversion. So delighted and relieved was Pedro Menéndez de Avilés by the scheme that he offered to supply a ship to carry the party to Bahía de Santa María.[10]

Menéndez and Don Luis reached Havana in November 1568 and soon met with Father Juan Baptista de Segura, head of the Jesuit mission, who had sailed over to Cuba from San Agustin. Despite the objections of nearly all of the other missionaries, Father Segura believed the American and accepted his offer to join the expedition to Ajacán in the capacity of both guide and interpreter. After a delay of many months, the ship, commanded by Vincent Gonzales, left Santa Elena and arrived "in the land of Don Luis" on September 10, 1570.[11]

Although Ajacán had been described by Don Luis—honestly, as he remembered it—as a most fruitful and beautiful country, the fathers were disconcerted to find that for several years both famine and disease had plagued the natives and that food was very scarce. The inhabitants, however, seemed very kind. Fathers Quirós and Segura wrote to their order on September 12 that this simple folk thought that "Don Luis had risen from the dead and come down from heaven." Two of his *cacique* brothers received the wanderer warmly, telling him of the death of an elder brother and that "a younger one was ruling." The latter was almost certainly the chieftain known to history as Powhatan.* He offered at once to yield all authority to his returned brother, but Don Luis, just as quickly, declined to accept the power and authority of the Powhatan, "asserting that he had not returned to his fatherland out of a desire of earthly things but to teach them the way to heaven which lay in instruction in the religion of Christ Our Lord. The natives heard this with little pleasure," the reverend chronicler reported.[12]

In a postscript to the letter just quoted, Father Quirós wrote: "Don Luis turned out as well as was hoped; he is most obedient to the wishes of Father [Segura] and shows respect to him, as also to the rest of us here." But this ideal state of affairs soon changed. Renewal of relations with his own people caused Don Luis to revert to old ways. A shocked Jesuit wrote that he "took to himself as many wives as the Gentiles," a behavior his tribesmen expected of one of their important chiefs.[13]

Severe reprimands sanctimoniously administered before the entire religious company, as well as delivered privately by

---

* *Powhatan* was the native place of Wa-hun-son-a-cock, who, when he became the principal chief (or emperor) of many Algonkian tribes in Tidewater Virginia, took the name of the village as his own. Succeeding chiefs were also known as *the Powhatan. Powhatan* was also the name of the empire of tribes that Powhatan formed. In this work *Powhatan* will be used in all three senses, the particular meaning being clear in context.

Father Segura and others, so humiliated the proud young man
that he renounced Christianity. In October 1570, he forsook
the Jesuits and went to live with his brothers in their village
on the Pamunkey River. Four months later, Father Quirós
and two other "religious" persuaded him that he should return
to the mission. Don Luis and some of his fellow tribesmen followed them on their way back and, on February 4, 1571,
killed them (Pl. 1), burned their bodies, and appropriated their
clothes and packs. Moving on to the mission, the Indians killed
the remainder of the Spaniards, except for one boy who ran
away and hid. These incidents naturally terminated all connections between the humiliated native American and the
Jesuits; hatred of all Spaniards and Christians took the place
of submission and respect. To signalize the transformation he
discarded the name Don Luis and took a new one, Opechancanough, which in Algonkian meant "He whose soul is
white."[14]

Scant though the evidence may be, we may surely infer
that Opechancanough, after 1571, spoke frequently and at
length with his brother about his experiences and observa-
tions during the decade he had spent in Spain, Mexico, and
the Caribbean with the white Europeans—matters that dis-
turbed and frightened Powhatan because they were so mysti-
fying and, seemingly, beyond his mental grasp. Such talks
serve to explain in part the myths periodically taught by their
"priests" that Opechancanough had come from the southwest
or the West Indies and the "prophecies" that sooner or later
the Powhatan tribes would be defeated, perhaps destroyed, by
invading strangers. At least from 1571 onward the brothers
were acutely aware of the menace from without, as well as
of the great strength of the white men, long before the English first appeared at Jamestown in 1607.[15]

Plate 1. Don Luis de Velasco killing the Jesuits, 1571.

Patently, it is more than a coincidence that in the years following 1571, what the English referred to later as Powhatan's empire was being put together by conquest. From six original tribes, the empire eventually expanded to thirty-two, and the system of autocratic control by *the Powhatan*, to whom all subject tribes paid tribute, was the most complex that any other Algonkian-speaking Indians ever managed to fashion. It seems clear, given the superior intelligence, knowledge, and experience of Opechancanough, that he was the principal architect rather than Powhatan. Time and again the returned native must have regretted his hasty decision not to accept the powerful office of *the Powhatan* when his naïve, and often irresolute, brother offered it to him.[16]

The policy determined upon by Opechancanough and accepted by the chieftains for dealing with white strangers was that, to save themselves, the Powhatans must exterminate them at some chosen moment. The idea was craftily perpetuated through the "prophecies" rehearsed periodically by their priests to keep the tribesmen informed and alert. Although largely unsuccessful, the two Spanish expeditions to the Chesapeake following the murder of the Jesuits in 1571—Opechancanough's *first* "massacre"—proved to the natives that their power and even tribal existence could be threatened by attack from outside their territory.[17]

The first recorded encounter of the English with the former Don Luis, now Opechancanough if our hypothesis is correct, took place late in May 1607, when Captain Christopher Newport's party was returning downstream from exploring the James River and stopped off to visit "the King of Pamaunches." Gabriel Archer wrote in his journal: "This king (sitting in the manner of the rest), so set his countenance, striving to be stately, as to our seeming he became [a] fool." The self-confident Britons never suspected that this apparently simple-minded chief was a man both sagacious and subtle who had been at the great court of Spain in the 1560s and possessed a remarkable understanding of Europeans and their ways. For the time being, Opechancanough was content to be taken for a "fool" while he artfully assessed the motives of these strangers.[18]

In December 1607, Captain John Smith and a party, seeking provisions, were surprised on the Chickahominy and captured by Opechancanough's warriors. Expecting to be killed, Smith attempted to buy time by giving the chief a compass and volubly explaining its use to him. The ruse won only an hour's delay, however, but for reasons of his own Opechancanough ordered that Smith's life be spared and had him taken before his brother, Powhatan, who also spared him and allowed him to return to Jamestown. To all appearances, relations were seemingly friendly between the Indians and the English.[19]

John Smith understood the natives better than his white companions did, but he failed completely to gauge the depth of the resentment of the English intrusion on the part of Powhatan, Opechancanough, and their fellow kings. This antagonism was greatly intensified in January 1608, when the Captain led a party of sixteen to the Pamunkey to persuade Opechancanough to trade for corn. When the werowance proved unwilling to do business and the Indians gathered in force, the fiery soldier boldly "did . . . take this murdering Opechancanough . . . by the long lock of his head," so he wrote in after years, "and with my pistol at his breast (Pl. 2) I led him [out of his house] amongst his greatest forces and before we parted made him [agree to fill] our bark with twenty tons of corn." Then the diminutive Englishman admonished the

C Smith taketh the King of Pamavnkee prisoner 1608

Plate 2. Captain John Smith threatens Opechancanough, 1608.

assembled Pamunkeys that if they did not promptly fill his ship with corn he would lade it with their "dead carcasses." Such a display of audacity, backed by European weapons, produced the desired result, but the public humiliation of their chief and the open threat to the entire tribe were never to be forgotten or forgiven by any of the Powhatan tribes.[20]

The kidnapping of Powhatan's "darling" daughter, Pocahontas, in the spring of 1613 by Captain Samuel Argall brought the already strained relations of the two races to the breaking point. This action may properly be considered the first of a long series of events that led directly to the quick decline of Powhatan's authority, the equally rapid rise of Opechancanough's fortunes, and onward to the tragic "Massacre" of 1622.

For almost a year Pocahontas was detained as a hostage in Jamestown. In exchange for the princess, Governor Sir Thomas Dale demanded that her father must return several runaway Englishmen, certain stolen tools and firearms, together with 500 bushels of corn and a guarantee "to conclude a firme peace for ever with us." These were harsh terms indeed, but Sir Thomas doggedly believed that the only other course was "present warre." Understandably Powhatan refused to meet John Rolfe and Robert Sparkes who, in March 1614, had come up the Pamunkey, bringing Pocahontas with them, to negotiate the peace.[21]

As the aged Powhatan's designated "successor" and already in "command of all the people" of the Powhatan empire, Opechancanough received the white envoys and promised to do everything he could to effect the exchange of persons and use his influence for peace. Before returning to Jamestown, the white men pointedly warned the chief that, if a final agreement were not made by harvest time, they would come again, kill the Indians, and destroy their houses and crops. Because the time was not yet ripe for resistance, Opechancanough apparently persuaded his brother to bow to the English ultimatum, which the Indians—quite properly in our modern view—looked upon as an imposed, personal truce never negotiated or written down but agreed to by the Powhatan "more from feare then for love." He also allowed Pocahontas to return to Jamestown and marry John Rolfe, as she ardently desired.[22]

Accompanied by two of Powhatan's sons, Opechancanough attended the marriage of John Rolfe and Pocahontas in the church at Jamestown, where he gave the bride away. This union did not produce the truce, but it did somewhat gloss over the manner in which the deed had been done—by a show of force. Sir Thomas and the rest of the colonists had won the return of the runaways, the restoration of the arms and tools, obtained an ample supply of corn, and, above all, had brought an end to "five yeeres intestine warre." To them the marriage was but "another knot to bind this peace the stronger."[23]

Opechancanough deftly turned the forced truce to the advantage of both himself and the Powhatan empire by inducing the strong and independent Chickahominy Indians to sue for peace with the English. In 1614 Samuel Argall concluded a written, formal treaty of friendship and alliance with this tribe, which also agreed to send a tribute payment of corn to Jamestown annually. Out of gratitude for his advice and assistance, these natives made Opechancanough their ruler—the "King of Ozinies." Here was a master stroke of forest diplomacy by which Opechancanough deluded the English into believing that the Chickahominies were their allies at the same time that he was secretly drawing the once recalcitrant tribe closer to membership in the Powhatan empire.[24]

Sir Thomas Dale, with the Rolfes and about ten Powhatans, sailed for England in April 1616. George Yeardley, a bluff soldier who was named deputy governor,* energetically promoted the culture of tobacco to the neglect of other crops, and soon the food supply of the colony fell dangerously low. Sending to the Chickahominies that summer for the tribute corn required by the treaty of 1614, he was given "a bad answer," one prompted on the sly by the new King of Ozinies. Completely unaware of the part played in the dispute by Opechancanough, Yeardley turned to him for advice and was told that the Chickahominies had been killing the colonists' swine and cattle. Janus-like, the Indian urged the Governor to march into the Chickahominies' country and seek a parley, while at the same time he craftily pressed his "subjects" to resist any English demands.[25]

Opechancanough completely outmaneuvered the English, who apparently never realized that fact, for they responded in a most injudicious and foolhardy way. Two or three hundred warriors gathered on the banks of the Chickahominy River to meet Governor Yeardley, who arrived with about a hundred armed men. Greeting him "with scorne and contempt," the Chickahominies, saying they had already paid Governor Dale, refused him any tribute. The next day, Yeardley, seething over his reception, ordered his men to

---

* When a governor did not arrive promptly or left the colony for England on a temporary leave during his term, or for good, he appointed a lieutenant or a deputy governor to take charge—both of them being addressed as governor. See Appendix IV of the present work for a list of the governors and deputy governors of Virginia, 1607-99.

fire upon the Indians, and between twenty and forty of them were slain. Some years later John Bargrave recalled that Yeardley's "perfidious act" caused all of the Chickahominies to "flye out and seek Revenge," and, far worse, it drove the tribe to make a common cause with the Powhatans, thereby freeing Opechancanough to form more alliances with other Algonkian tribes of Tidewater Virginia.[26]

In the meantime Powhatan, old, tired, and ineffective, was steadily losing control of his empire. Suspecting Opechancanough of conspiring with the English against him, he decided to alter the succession and designated Itopatin, a lame younger brother, the next emperor. In 1617 he delegated his authority to Opechancanough and Itopatin jointly, but on his death in April 1618 the latter became *the Powhatan*. Thereupon a contest for leadership of the tribes ensued between Itopatin and the more popular Opechancanough, from which the latter emerged victorious. He was ready this time to deal with the English situation.[27]

Several incidents occurred in 1619 that created tension between the red men and the white, but Sir George Yeardley, now governor, managed in each case to avert open warfare. On November 11 he asked the advice of his Council about a request from Opechancanough for eight or ten armed Englishmen to assist him in punishing some natives beyond the falls of the James for killing some Pamunkey women. In return, the chief offered half of the plunder of corn, lands, and children of both sexes. Finding the war justifiable, lawful, and inexpensive, the councilors agreed that it would be well to placate Opechancanough, whose dislike of Yeardley had been evident since the slaughter of the Chickahominies. They hoped also to win the "amity and Confidence" of Itopatin and the other Powhatans. But especially did they welcome the chance to procure some Indian children for the projected school at Henrico. The requested aid was therefore given.[28]

A new administration of the Virginia Company of London, managed by Sir Edwin Sandys, attempted to reinvigorate the colony by a grant of self-government to the inhabitants and the sending over of large numbers of new settlers. Grants of great estates to certain favored Englishmen and fifty acres of land to each person transported who stayed three years (a headright) had presaged an expansion of the English after 1618 into the Indian lands along the Chickahominy and upper James rivers. There was actually little encroachment on the natives before the end of 1622, but officials in London proposed in June 1619 that, upon his return to Virginia, Yeardley should begin "the removal" of the Chickahominies "by all lawfull meanes" in retaliation for several murders of whites recently committed by them in "revenge" for Yeardley's massacre of 1616. Opechancanough probably never learned about this projected course. He did, however, know all about English intentions to assimilate his people by conversion to Christianity, the religious education of their children, forced labor, and other devices, and he could have recognized the resemblance to the policy of "congregating" the Indians in villages that he had observed in Mexico years earlier. At any rate he steadfastly refused to allow their children to be brought up in the English faith, language, and manners.[29]

The Virginia Company under Sandys's energetic leadership implemented the provisions of the Charter of 1606 for converting the natives. As the Reverend William Crashaw had put it with unconscious irony in 1609: "Out of our humanitie and conscience, we will give them . . . such things as they wante and neede, and are infinitely more excellent then all we take from them . . . 1. Civilitie for their bodies, 2. Christianitie for their soules; the first to make them *men;* the second *happy men.*" In Virginia, however, both officials and people not only neglected the work of conversion but looked down upon the natives with ill-concealed contempt. Almost to a man they ignored a well-deserved reproof by Alexander Whitaker: "Let us not think that these men are so simple as some have supposed them: for they are of body lusty, strong, and nimble; they are [an] average understanding generation, quick of apprehension, sudden in their dispatches, subtle in their dealings, exquisite in their inventions, and industrious in their labour."[30]

Whatever suspicions John Rolfe and other local officials may have entertained about Opechancanough in the first decade of the colony's existence, during 1620 they were writing to London of the confidence in "their assumed peace" and that "every man planteth himself where he pleaseth and followeth his business securely." Not far from Jamestown, the great chieftain, still secretly and quietly working at backwoods diplomacy, was using every means at his command to instill a sense of security in the colonists. So safe had the English come to feel and so convinced were they that they were conferring great benefits upon the Indians that they failed to recognize that the animosity of the natives for the white men had grown rather than diminished. This enmity, which was due to a quarter of a century of insults and mistreatment, was now being exacerbated by fear of the English encroachments on their lands.[31]

Opechancanough's distrust of the Englishmen's professions of good will did not deter him from conducting a brisk traffic in corn and other commodities with the little community of Jamestown during the "four years of security" after 1618. His

profits enabled him to employ many "auxiliaries" from Algonkian tribes in the Tidewater that were not members of his
empire, something that Powhatan had never been able to do.
By this means and other achievements resulting from a covert
tribal diplomacy, his confidence, along with military strength,
mounted significantly. Nevertheless the accelerating influx of
settlers from England alarmed him—after 1619 the white population had risen from 1000 to 2300 and the potential in-
fringement on tribal lands, even more than the actual settlement, threatened the entire native society.[32]

Sometime during the spring or summer of 1621, Opechancanough became convinced that the time had come to strike
at the English; if they were not wiped out, they could soon
overwhelm and destroy his people. He felt strong enough
now to make a surprise attack if and when a favorable opportunity arose. Rumors spread about that the Indians were plot-
ting to fall upon the English in September or October, and
the colonists were warned to prepare for it. Opechancanough,
learning of this, earnestly and vigorously denied the possibility
of any such attack, and shortly the settlers "fell againe to their
Ordinary watch." In November, Sir Francis Wyatt arrived,
replacing Sir George Yeardley as governor, and reported that
the country was quiet and "all men in a sense of security."
Not only did Opechancanough prove willing to renew "the
League" of peace, which he had refused to do for Yeardley,
but surprisingly suggested that some white families live among
his own tribesmen.[33]

It was a man "farr out of the favor of Apachankano" who
unknowingly provided the opening for the showdown of the
Powhatans with the English. Shortly before the discovery of
the Indian plot scheduled for the fall of 1621, Nemattanow,
called Jack of the Feathers by the white men, had killed sev-

eral settlers. Early in March 1622, two servants of one Morgan, in retaliation for the murder of their master, shot and fatally wounded the warrior. Jack was a famous fighter of little sense or judgment, and when the news of his death reached Opechancanough, he notified Governor Wyatt at once that Nemattanow "beinge but one man, should be no occasion of the breach of the peace, but *that the Skye should sooner falle then [the] Peace be broken,* one his parte, and that he had given order to all his People to give no offense and desired the like from us."[34]

In spite of these peaceful overtures, the shooting of Jack of the Feathers was precisely the kind of provocation that the Powhatan had been waiting for to launch an all-out attack against the white peril. This was the goal he had hoped and planned for since 1613, and he believed it was now or never. Jack was highly regarded by the tribesmen as one of their greatest braves, and though Opechancanough himself cared little for the man, he used the murder to stir up all of his race in Tidewater Virginia. The incident of Jack of the Feathers' death was only the excuse for, not the cause of, the Powhatan uprising.[35]

On March 22, 1622, the sky did fall. Everywhere in the colony, the native Americans attacked the Europeans and butchered 347 men, women, and children, including even those of the English who had treated them fairly and well. Opechancanough's plan called for the extermination of all of them, and he would have succeeded "if God had not put it into the heart" of a Pamunkey servant named Chanco to disclose the plot to his master Richard Pace "who had used him like a son." Pace immediately rowed across the James to warn Governor Wyatt, thereby saving the lives of the inhabitants of Jamestown and nearby plantations.[36]

Plate 3. The Massacre of 1622.

This deliberate, admirably timed attempt to exterminate the white invaders was, no doubt, thought of by Opechancanough and his followers as a stroke for Indian freedom. We have seen that since 1613 he had been planning an all-out attack. By means of diplomacy of which any European could have been proud, this chieftain united the Powhatans, their "auxiliaries," and the Chickahominies, worked out tactics with consummate skill, and decided upon the day and hour when, simultaneously, the natives were to fall upon all of the whites. The preservation of secrecy throughout the entire undertaking was miraculous. Employing both Indian and European methods of deception and surprise, he won a victory that elicited reluctant praise from many leaders across the sea. From Kent in 1624, George Wyatt wrote to his son, the Governor, pointing out that the Indians' "Intelligences served them wel[l]," and approved their organization and manner of fighting as the best possible for attaining their ends. All things considered, the "Massacre" of 1622 was probably the most brilliantly conceived, planned, and executed uprising against white aggression in the history of the American Indians.[37]

For several months Opechancanough apparently overestimated the completeness of his attack. Toward the end of March 1622, he urged the King of the Potomacs to wipe out a trading party on his river, assuring him that "before the end of two Moons there should not be an Englishman in all their Countries." Furthermore he underestimated the shame and resentment felt by the English at their defeat and their implacable determination to avenge fully what they deemed the Indians' "treachery." Although the "Massacre" was followed by an epidemic sickness that carried off double the number of settlers killed by the natives, recovery came promptly and permanently. War was declared upon the Powhatans in August 1622, and a price was set upon Opechancanough's head. By the end of the year the Indians were admitting that since the "Massacre" more of their number had been killed than in the entire period 1607-21.[38]

Despite the rejection of proposals for wiping out the tribes, the retaliation of the English was fierce, grim, and relentless. They set upon the natives "in all places," systematically destroyed their villages, and burned or cut down their corn and other crops, killing as many as they could. In the words of a ballad written in the colony and hawked about London in 1623, one English captain appeared at an Indian village "with honor":[39]

> Who comming took not all their corne,
> but likewise tooke theire King
> And unto James his Citty he,
> did these rich trophies bring.

Such spoils as the English seized they used to feed the colonists who had survived the "Massacre" and fled to Jamestown. Two-thirds of these refugees were "women, children, and unservisable persons." But in spite of fear, fevers, and famine, the local leaders could not be persuaded to accept the "ignominious proposician of removeinge to the Easterne Shore" of Virginia for safety. The members of the Council advised the Company in January 1623 that "they had Carefully repared the decays of James Citties, and invited all men to builde theire," and that reconstruction was "proceeding Chearfully one." Never again would the tiny metropolis of Virginia be menaced by Indian attack.[40]

From August 1622 until 1632, with one brief interval of uncertain peace, the white men's war of revenge against Opechancanough, Itopatin, and the Powhatan tribes continued.

The Indians fought desperately, but in vain. They were forced out of the peninsula between the James and York rivers into lands west of the Fall Line; more of them died from starvation than at the hands of their English foes. But the final act of the first lost cause took place on the streets of "James Cittie" in 1644, when the centenarian Opechancanough, who at long last had been taken captive, was shot in the back by an English soldier.[41]

III

John Rolfe, Pocahontas,
and "that chopping herbe of Hell,"
Tobacco

From the very outset, the officers of the Virginia Company intended that the settlers in their proposed colony should produce their own food, trade with the natives, and send back home a variety of vendible commodities, some exotic, others of their own manufacture. Virginia was a country where fruits, berries, game, and fish existed in profusion, and to the organizers of the colony their expectations did not seem unreasonable. Present-day hindsight makes clear that the Englishmen who first went out to the colony were an unskilled, improvident, and lazy lot who, "no more sensible than beasts, would rather starve in idleness . . . than feast in labor." Like the grasshoppers in Aesop's fable, they took little if any thought for the future, and as a result, many of them perished, not only during the "starving time" but also for years to come because of their failure to plant crops.[1]

Powhatan and his people looked upon the strangers' intrusion with fear and distrust. Nevertheless they were willing, at first, to teach the Englishmen how to plant beans and squashes, how to catch fish in weirs, and how to hunt wild turkey and deer. They also introduced them to the two most important crops of the New World—maize (corn) and tobacco. The men from Britain failed to plant corn; they preferred to get it from the Indians either by trade or by force—Captain John Smith proved himself adept at both methods.

Tobacco was an exotic product that was destined to fill the description of a "vendible commodity," which had figured in the original planning. When the first settlers moored overnight at Kecoughtan before entering the James River, George Percy tells us that an Indian who had "a garden of tobacco" generously "distributed some to every one of us." Those colonists who had smoked tobacco in England found the local leaf "poor and weak, and of a biting taste." They much preferred the Spanish tobacco from Trinidad or Caracas, and soon seeds of these varieties were imported from the Caribbean and South America and planted in Virginia. The results were meager until about 1611 when one of the settlers began to experiment with seed from Trinidad, "partly from the love he hath for a long time borne unto it, and partly to raise [a] commodity" to sell in the British Isles.[2]

The first colonist to succeed in growing marketable West Indian tobacco was John Rolfe, formerly of Norfolk in East Anglia, who had reached Virginia from the Bermudas in 1610. His ship, *Sea Venture*, was blown aground there in a storm on the voyage out to Virginia, and several months passed before the survivors could build two pinnaces from the timbers of the wrecked vessel and proceed to the Chesapeake. During this delay, Mistress Rolfe gave birth to a daughter, christened Bermuda, who soon died; the mother died shortly after they reached the colony. No man among the early English colonists of Virginia, not even Captain John Smith, contributed more, ultimately, to making the plantation a going concern or was so influential in giving direction to its destiny than the young widower.[3]

On April 13, 1613, at the very time John Rolfe was raising his first good crop of tobacco, Captain Samuel Argall brought Pocahontas, the favorite daughter of Powhatan and niece of Opechancanough, to Jamestown as a prisoner and hostage. This "well featured but wanton [lively]" friend of Captain John Smith, called Matoaka by her own people, had not visited her English acquaintances since she saved the soldier's life five years before. Now she had matured and blossomed into an attractive young woman of eighteen or nineteen years of age. According to a report written two years before her death by Ralph Hamor, an intimate of Rolfe, the young planter had fallen in love with Pocahontas within a few months of her arrival in the English village, "and she with him."[4]

There still exists a very touching letter to Sir Thomas Dale in which the pronounced Calvinist asks for the Governor's permission to make the heathen woman his wife. Rolfe had repeatedly examined his conscience to assure himself that to espouse a creature "whose education hath been rude, her manners barbarous, her generation accursed, and so different, discrepant in all nurtriture from my selfe," would be acceptable to God and for the good of the colony. Most convincing of all arguments, one suspects, was Pocahontas herself, "to whom my heart is and best thoughts are, and have a long time bin so intangled."[5]

Permission for the English gentleman to marry the Indian princess was quickly granted by Sir Thomas Dale—not the least of his reasons being the tenuous peace recently forced by the English upon Powhatan and his tribesmen. At the Governor's request, the Reverend Alexander Whitaker had come

down from Henrico to instruct Pocahontas in Christian prin-
ciples and also in the English tongue. She took readily to both
and soon renounced idolatry, openly confessed her Christian
faith, and, "as she desired," was baptized Rebecca. Although
her father, the aged chieftain, had but recently refused to
admit Rolfe to his presence, he favored the match for much
the same reason as the Governor did, and he sent two of his
sons, along with his brother Opechancanough [Apachisto],
to be present at the wedding. The uncle gave the bride away
in the ceremony that was celebrated in the timber church of
Jamestown on April 5, 1614, by the Reverend Richard Buck.
This interracial marriage did not end hostilities, as we have
noted, but there is no doubt of its relieving some of the tension
between the Indians and the English for some years.[6]

John Rolfe undoubtedly was a resident of Jamestown in
these years; his meeting with and courtship of Pocahontas
must have taken place there, for she was a hostage in the village during the years 1613-14. Furthermore since his arrival in
the colony, Rolfe had become prominent in governmental
activities: he was one of two men sent to negotiate with Powhatan in 1614, and, after his return from England in 1617, he
entered upon five years of service as the first secretary and
recorder general of Virginia, which required his presence in
the village. It seems only reasonable that he would have
planted his first tobacco on Jamestown Island and not at
Henrico as has often been surmised. In March 1614 just before
his wedding, he shipped off in the *Elizabeth*, bound for London, four barrels of the leaf he had grown in the past year.
The vessel reached port, the barrels were entered at customs,
and in July 1614 the fateful tobacco trade between the colony
and England had its inception. In the words of Philip Alexan-

der Bruce, this "was by far the most momentous fact in the history of Virginia in the seventeenth century."[7]

When Governor Dale was recalled to England in May 1616, he invited the Rolfes to sail with him in the *Treasurer*. After their arrival in London, John Rolfe wrote a report for perusal by King James I, the Earl of Pembroke, and Sir Nathaniel Rich entitled *A True Relation of the State of Virginia, Left by Sir Thomas Dale, Knight, in May last, 1616*. As one might expect, the secretary put the most favorable construction upon affairs in the colony, but nevertheless the tract was full of valuable information; it also shows that its author wrote in a clear, and surprisingly puritan, spirit, and that he could also be entertaining.[8]

Ten or twelve other Indians besides Rebecca Rolfe were among the company traveling with the Governor, and all of them excited the curiosity of the London gentry. It was Mistress Rolfe, however, now able to speak English with some fluency, who charmed them all. The Reverend Samuel Purchas, to whom John Rolfe had lent his report, *A True Relation of the State of Virginia*, had been introduced to her at an entertainment given by the bishop of London and was visibly impressed. In his collection of travels, *Purchas His Pilgrimes* (1625), the clergyman remarked that she "did not only accustome her selfe to civilitie, but still carried her selfe as the Daughter of a King, and was accordingly respected, not onely by the Company . . . but of divers particular persons of Honor in their hopefull zeale by her to advance Christianitie." There must have been many who were saddened when they learned that, less than a year after her arrival in England, while awaiting favorable winds for the return to Virginia, Pocahontas unexpectedly sickened (probably of tuberculosis) and died at Gravesend, March 21, 1617. She was buried there

MATOAKA ALS REBECCA FILIA POTENTISS PRINC: POWHATANI IMP:VIRGINIÆ.

Ætatis suæ 21. A̴
1616.

Matoaks als Rebecka daughter to the mighty Prince Powhatan Emperour of Attanoughkomouck als virginia converted and baptized in the Christian faith, and wife to the worth. Mr. Joh Rolff.

Plate 4. Pocahontas in 1616.

in the parish church. Her husband, after placing their son Thomas with a relative, sailed, in April, to resume life at Jamestown.[9]

For two years after his return, Rolfe continued to serve as secretary and recorder of the colony, but he relinquished these offices when he was admitted to the Council in 1619. He participated in the first legislative Assembly, and in 1621 he was designated, by the Company, a member of the new Council of State. Sometime during this period, Rolfe married Jane, the daughter of Captain William Pierce of Jamestown. Within a year, however, before the "Massacre," he died from natural causes, apparently in Jamestown. Putting aside all the romantic tales about John Rolfe and Pocahontas, "he remains," his biographer tells us, "one of the great heroes of our colonial era."[10]

John Rolfe was the first planter on record to domesticate West Indian tobacco and initiate the tobacco trade with the mother country—and as late as 1621 he was still shipping barrels of it to England. Very shortly other planters, sensing the opportunity to profit by its production and sale, began to emulate him. Ralph Hamor had insisted in 1612 that "I dare thus much affirme . . . of Tobacco, whose goodnesse mine own experience and triall induced me to be such, that no country under the Sunne, may, or doth affoord more pleasant, sweet, and strong Tobacco, then I have tasted there; even of mine owne planting, which, howsoever, being then the first yeer of our triall thereof, we had not the knowledge to cure, and make up."[11]

It was George Yeardley, the successor to Dale as deputy governor in April 1616, who should be credited with turning the weed into the prime staple of the Old Dominion. He urged the planting of it to such good effect that, when Samuel Argall came to Jamestown in 1617, he found to his dismay "the market-place, and streets, and all other spare places planted with tobacco." Before long, everywhere up and down the valley of the James, the settlers were growing and sending it home to be sold. The ships took away 2,500 pounds of tobacco in 1616; 18,839 in 1617, 49,668 in 1618; and in 1628, the remarkable total of 552,871 pounds arrived in the mother country from its first American colony.[12]

The Company at London had never approved of restricting agriculture in the colony to a single crop; in fact, Samuel Argall's reports about the extent of the tobacco-growing craze greatly alarmed the members. Sir Edwin Sandys warned the governor, Sir George Yeardley, in 1621, against the settlers' overplanting the seed and neglecting to grow corn to feed themselves, but it was too late to diversify the economy of Virginia. All attempts were doomed to failure for, as the Earl of Warwick bluntly put it, the planters abandoned "all other staple commodities" because of their "overweening esteem of their darling tobacco."[13]

Many of the refugees from the Indian fury, who brought their "cattle"—kine, hogs, goats, horses—along with them when they came seeking asylum in Jamestown in 1622, discovered that there were some advantages in living on the island. For one thing, the beasts were safe from predatory animals, and it also became apparent that the swamp and marshland afforded pasturage. Moreover, grazing required less backbreaking toil and fewer laborers than raising tobacco; and near at hand at the waterside where local and English vessels moored, was a ready market for salted meats, hides, and tallow. All efforts to develop arts and crafts—ironworks, glassworks, and other industrial enterprises—were short-lived; they simply did not yield the profits that were to be gained from tobacco.[14]

One disturbing consequence of the influx of planters to Jamestown after the "Massacre," which became apparent very quickly, was that land for growing tobacco on the island was very limited, and, furthermore, the soil was already noticeably exhausted from raising the same crop year after year. More and more after 1622, the gains to be made at Jamestown from the culture of the "noxious weed" came from related pursuits.

In the decades following the "Massacre," a small but exceedingly prosperous and powerful group of merchant-planters and factors (from London and Bristol houses) arose, and many of its activities centered at the little village when it became the chief port of the James River trade. The courts and legislature were located there, and as we turn the pages of the journals of the House of Burgesses or the Council of State, we perceive at once that the principal business of these worthies was tobacco. In 1639, for instance, the lawmakers provided for sworn overseers to inspect the quality and weight of the staple. The next year Governor Wyatt decided it was necessary to issue a proclamation from "James Cittie" forbidding the cultivating of more than a thousand plants by each farmer and ordering the planting of more corn. Sir William Berkeley wrote acidulously in *A Discourse and View of Virginia* (1663): "The vicious, ruinous plant of Tobacco I would not name, but that it brings more money to the Crown, then all the [West Indian] Islands in America besides." He blamed the economic plight of the colony upon a tobacco culture that was founded upon "the vices of men."[15]

David Pietersen de Vries, Dutch skipper and trader who was in the James River region collecting tobacco debts in

1644, observed: "I occasionally examined their plantations, and found that lands which have been exhausted by the to-bacco-planting were now sown with fine wheat, and some of them with flax." This was the last resort of men who had worn out their lands by growing tobacco exclusively. When-ever and wherever possible, most planters, large and small, patented and cleared more virgin land, leaving the exhausted "old fields" vacant for lesser fry, who had to compete with the many new immigrants who were arriving each year and wanting land. The pressure for additional territory was the principal cause of the last futile Indian uprising of 1644. Two years later Governor Sir William Berkeley dictated to the new Powhatan, Necotowance, a treaty whereby all of "the peninsula" between the James and the York, as far inland as the falls, was ceded to the English. Before long, land-hungry tobacco planters were pushing northward beyond the York to the Rappahannock. Tobacco had become king.[16]

The Indians had given the English in Virginia a staple by which they were enabled to establish and maintain a lucrative commercial connection with the outside world. This is a prime need of any new colony, anywhere, any time. Tobacco, the miraculous staple pungently labeled by Thomas Dekker "that chopping herbe of Hell," created as many problems for Vir-ginia—later for America—as it solved. Grown first at James-town by John Rolfe, Ralph Hamor, and their friends, and spreading out from there, it not only dominated the agricul-tural economy of the Old Dominion but led to the one-crop farming that eventually plagued the entire South until our own day—tobacco, rice, indigo, sugar, cotton. For this Amer-ica paid a terrible price. This development began at James-town, and it took its toll from the red men, the white men, and the blacks.

IV
Life and Death at Jamestown

i. The Cost in Individual Lives

After a spirited debate on November 16, 1667, in the House of Commons about the draining off of Englishmen by the plantations, Mr. Garroway, a member of that body, remarked privately that "the consumption of people" by Virginia was so great that the colony needed a new supply of settlers annually in order to survive. Almost in a state of shock, the member told his listeners that "the English at their first arrival lose a third part at least." It should be remembered that he was talking at a time when mortality had noticeably declined and, of course, was referring only to white Britons. The incredibly high cost in human life paid to establish the settlement from 1607 onward should include not only the lives of white men but also of red men; and after 1680 the lives of black men too. But let us begin with the English of whom Mr. Garroway spoke so feelingly.[1]

The loss of life at Jamestown and its environs in the first two decades defies the imagination. A greater number of Englishmen were buried on the island than lived there at any one time. For the most part these unfortunate creatures were hur-

44

riedly interred without ceremony or coffins lest the natives discover the truth about the incidence of death at the outpost. From the departure of the first ships from England in December 1606 to February 1624—the years of the Company's rule— out of 7,289 immigrants, 6,040 succumbed, or six died for every one that lived. A cemetery with unmarked graves would be their only memorial.[2]

Writing about the first settlers, George Percy declared: "Our men were destroyed with cruell diseases, as Swellings, Flixes, Burning fevers, and by Warres [with Indians], and some departed suddenly, but for the most part they died of meere famine. There were never Englishmen left in a forreine Countrey in such miserie as wee were in this new discovered Virginia." During its entire history, the Virginia Company's officials appear never to have thought it was their duty to see that their colony was properly supplied with food. When Ralph Hamor and Sir Thomas Gates "sailed sadly up the river" to Jamestown in the spring of 1610, right at the end of the "starving time," they found not more than "three score persons therein, and those scarce able to goe [it] alone, of welnigh six hundred, not full ten months before"—a ratio of 1 to 10.[3]

Contemporaries were quick to point out that many of the basic causes of this dreadful mortality originated in the British Isles. An overwhelmingly majority of those people from overseas, whether English, Irish, or African, who landed at Jamestown and spent a brief time there as transients before 1699, was in very bad physical condition on arrival. Before they boarded the ships at London, Southampton, or Bristol, many of the emigrants were suffering from malnutrition, some with jail fever, and others with various communicable diseases.

Conditions on shipboard during the Atlantic crossing can be

charged with adding to the high death rate among the emigrants who set out for the New World with hopes of a better life ahead. William Capps of Kecoughtan pointed out in 1623 that the chief cause of unhealthiness at sea was the want of cleanliness, which could have been prevented. When he crossed with Sir George Somers in fifteen weeks in 1609, no one was lost, but in most ships, however, "betwixt the decks there can hardlie a man fetch his breath by reason there arises such a Funke in the night that it causes putrefaction of the bloud and breeds a disease much like the plague; the more fall sick, the more they annoy and poyson their Felloes." Others, he insisted, were sickened by "the Poyson they gave us instead of Beare." Because the emigrants represented cargo rather than passengers, unscrupulous merchants and ships' masters packed them in unmercifully—"pestered" was the word of the day. This overcrowding of ships, which "carry with them almost a general mortality," was the unpardonable crime of the merchants.[4]

During the long journey across the ocean, which lasted anywhere from six weeks to several months, poor and insufficient food, foul water, cold, and dampness all contributed to debilitating an entire ship's company, and deaths at sea were common occurrences. In 1636 Governor John West complained to the Lords Commissioners of Trade and Plantations that whenever a parcel of immigrants debarked at Jamestown, most of them were in a weakened state and some were carrying diseases easily transmitted to the colonists or to the singularly unresistant natives.[5]

In Virginia all newcomers underwent a year of "seasoning" that proved fatal to large numbers of them. David Pietersen de Vries thought it a fine country in 1633, but he observed that the English had a serious objection to it: "They say that during the months of June, July, and August, it is very unhealthy; that their people who have lately arrived from England, die during these months, like cats and dogs, whence they call it the sickly season. When they have this sickness they want to sleep all the time, but they must be prevented sleeping by force, as they die if they get asleep." In 1648 Sir Edmund Plowden rejected Virginia as a place to settle because, among other objections, the salt marshes and creeks were "thrice worse than Essex, [the Isle of] Thanet, and Kent for agues and diseases, brackish water to drink and use, and standing waters in the woods that bred a double corrupt air, so the elements corrupted; no wonder, as the old Virginians affirm."[6]

For reasons not now entirely clear, the incidence of "seasoning" began to decline after 1650 and by 1699 was much reduced, but this sickness had killed 10,000 colonists by 1657; "generally five of six imported died." The mass of the men fresh off the ships, because of their debilitated condition, was especially susceptible but even physically sound persons died during the summer months. As late as 1663 only 20 percent of the newly arrived immigrants survived the first year.[7]

At any given time, Jamestown, as a transfer point, had a large number of temporary residents among its population: white servants from England and Ireland, and later black slaves awaiting purchase by tobacco planters; captains of ships and their crews temporarily ashore (though many vessels remained for several months); and various traders. None of these people adjusted to the climate. Furthermore, the half-urban conditions and location of the small port were not conducive to good health: poor housing, the generally unsanitary situation on the island, the bad air rising from the marshlands and swamps, the plague of mosquitoes, and the water that caused typhoid fever. In addition there were innumerable rats, which in 1616 not only consumed nearly all of the settlers' corn but spread disease every year. Whenever a ship brought in a passenger with a communicable disease, some Jamestonian seemed bound to contract it. In the 1680s, the Reverend John Clayton, who was trained in medicine, pointed out the prevalence of many ailments in his parish.[8]

Inability to adjust to a ,new country, discouragements and despair, fear of the Indians, strange ways, unaccustomed to toiling in hot climate, and misrepresentations, as well as a host of psychosomatic difficulties of which we know all too little, must have contributed heavily to the great loss of life at Jamestown. George Thorpe was persuaded that "more doe die here of the disease of their minde then of their body by having this countrey victualles over-praised unto them in England and by not knowinge they shall drinke water here." Being predominantly male, the townsfolk had no family life; nearly all of them were lonely if not homesick. The ruthless and often heartless treatment of servants by masters capped this somber existence for the majority.[9]

The will to live was not strong in such defeated men. In a discussion of the Virginia climate in 1623, Governor Wyatt commented that the "weaker sexe also escape better then the men" either because women worked chiefly indoors or because they were of "a colder temper." According to a sort of travel guide to the New World published in 1651, "The Virginia proverb is, *That hogs and women thrive well amongst them.*" This writer recognized that females resisted seasoning and disease better than males; if they survived childbirth, their longevity, as in England, was noticeably greater.[10]

The greatest loss of life in Virginia probably occurred between 1621 and 1623 when, according to Company estimates, the deaths of white colonists numbered around three thousand, only three to four hundred of whom were killed during the "cruel tragedy of the massacre." A far greater number (86 percent) succumbed to infection from the ships or scarcity and despair in the colony. George Baldwin told a gruesome tale to a friend in Bermuda: "It hath been a very harde time with all men; they had like to starve this yeare. There was them that paid fortye shillings a bushell for sheld corne. But howsoever, they dye like rotten sheepe; no man dies but he is as full of maggots as he can hold. They rott above ground." As for the country itself: "I like it well if the people were good that are in it, but they are base all over, for if a man be sicke, put him into a new house, and there lett him lie down, and starve, for nobody will come at him."[11]

Although the Indians did not succeed in slaying anyone at Jamestown in their all-out attack, in the following winter, 1623, more people died than had been slaughtered by Opechancanough's braves the previous summer. The influx of refugees from the surrounding countryside, the loss of crops, malnutrition, famine, and the uneasy peace, all combined, explain most of the deaths in the island. The report was that there were not many of its inhabitants "that have not knockt at the doores of death."[12]

The colonizing of Virginia by the English exacted a cost in human lives from the natives that was as great, possibly greater, than that from the white men, but we seldom see reference to this grim truth. As we noted earlier, the only remaining sources are those left by white men, but this, in a sense, lends more credence to what we learn about the Englishmen's treatment of the Indians. They were indisputably responsible for the deaths of thousands of the aboriginal population.

When the settlers first arrived, the natives were practicing a sedentary agriculture around their villages and living in great plenty. They were accustomed to an ample supply of food and enjoyed a considerable variety in their diet. They also regularly ate more meat than the rank and file in England, which explains, in part, their stature and physiques. Unwilling to farm for themselves, or unable to do so, the invaders demanded a share of the "savages' " corn and provisions, and later settlers began to fan out from Jamestown and appropriate tribal lands. When, in defense of their own lands and property, Opechancanough, Powhatan, and other werowances dared to resist these aggressive strangers, the English warred upon them with superior weapons and, at the same time, irrationally destroyed the supply of grain and fish sorely needed by both races.

The General Assembly of Virginia admitted in 1621 that from 1609 to 1614 "we were at war with the natives," and when any Englishmen were slain, Sir Thomas Dale had revenged their blood "by divers and sundry executions, in killing, cutting downe, and takeinge away their corne, burning their houses, and spoiling weares, etc." In the times of wavering peace, 1614-21, this formula was not abandoned, rather it was regularized. Contagious diseases common among the white men spread to the Indians, cutting down their ranks; in the torrid summer of 1619, both the Englishmen and the tribesmen were visited by a great sickness and mortality. In return, the native women infected the lascivious English and Irish men with the malady "called the Country Duties," which they treated as they did "the French Disease, it being almost alike." It was probably the yaws.[13]

The initial shock of not having sufficient food for themselves after 1614 propelled the Indians into fighting back as best they could against the new enemy's greed and ruthlessness. They came very close to exterminating the invaders in 1622, but the odds were against them in the total war launched by the white men after the "Massacre." More devastating than the casualties suffered in open conflict were those resulting from the planned diabolical scheme of driving them from the neighborhood of Jamestown by starving them. The colonists persuaded the "savages" to return to their villages in 1623 and resume their old practice of planting their crops in adjacent fields; when the corn was ripe, the English would swoop down and destroy the crops and burn the houses; and they repeated this unchristian performance twice a year. More Indians were starved to death than were driven out of the James valley by 1646. (Few readers of this history will realize that the first and oldest Indian reservation of the present United States was established on the Pamunkey River in 1653.)[14]

After two conspicuous failures at winning freedom, the Powhatans, by 1650, were completely defeated, and by 1699 the remnants of those once proud and powerful tribes had been forced westward beyond the limits of the Tidewater. Theirs is a tragic story: they paid the supreme price of extermination, which, we must point out, more than doubled the human cost of the founding of Jamestown.

The first blacks known to have been sold into some kind of servitude in Virginia were brought to Point Comfort by Captain Jope in a Dutch ship guided by Master Marmaduke (Rayner?), an English pilot, in August 1619. Governor Yeardley and the "cape Merchant," Abraham Piersey, bought them, paying for them with provisions. Apparently the "twenty and odd Negroes" were taken upstream to Jamestown where their new owners lived, for in a census of 1625 ten Negroes were listed for Jamestown Island. A year later, Yeardley and Piersey had twelve blacks, and elsewhere on the island there were eight. Whether the Africans, probably brought from the West Indies, were servants or chattel slaves cannot now be ascertained. In 1640 a white man got a female Negro servant pregnant and was ordered, by the court, to "do public penance for his offence at James City church in the time of divine service, according to the laws of England in that case provided." The Negress was whipped, as a white woman would have been, but the implied expectation of Christian morality in blacks was certainly not characteristic of later slaveholders.[15]

George Menefie, a merchant of Jamestown and a member of the Council, received in 1638 a patent for 3000 acres of land for sixty persons whom he had imported; of these he described twenty-three as "Negroes I brought out of England with me." Being a dealer in white servants, he probably disposed of the blacks by a public sale in Jamestown, either as servants or, considering the date, as chattel slaves. The village, thereby, would have been the scene of the first slave market in America by at least 1638.[16]

From 1640 onward, the status of blacks slowly evolved from servitude to slavery as the General Assembly progressively defined the "peculiar institution." That there ever were many black slaves living in the village itself during the seventeenth century may well be questioned, for there existed little work for them; and on Jamestown Island grazing required but a few herdsmen. We know next to nothing about the blacks of the community and are therefore tantalized by the lawsuit of "Edward Lloyd, Planter," a mulatto of James City County who, in 1677, sued Captain William Hartwell of Jamestown for imprisoning him for three weeks as part of the retaliation after Bacon's Rebellion. During this time, Lloyd charged, Governor Berkeley's servants plundered his house and so frightened his wife, then "great with child," that she died as a result. Two white midwives, Mary Colby and Mary Robinson, deposed in his favor under oath, but the record is silent about the outcome.[17]

ii. The Cost to Society

So far we have been examining the high price paid in individual lives for the settlement of Virginia. Concurrently, groups of human beings had to pay, collectively, great social costs brought on by the transformation of "English Liberty" in the New World: a few gained but the many actually suffered a loss of hereditary freedom. Human abasement did not begin at Jamestown in 1619 when the first blacks arrived; it commenced in 1607 with the landing of the first white colonists from the *Susan Constant*, the *Godspeed*, and the *Discovery*. For nearly two decades (1619-38) this outpost of empire served as a market for the sale of white Englishmen and Irishmen into servitude before it became a mart for black slaves.

Until the loss of the charter in 1624, most of the colonists at Jamestown were, individually, servants of the Virginia Company of London. The reason for this state of affairs was that all but a very few emigrants from the British Isles to the Chesapeake colony in this century were too poor to pay for their passage. They were, however, *free persons*. Merchants or ships' captains borrowed from the English apprentice system the familiar device of the *indenture*—a voluntary contract stipulating that in return for passage, food, and clothing, the freeman would agree to work as a servant for a planter in Virginia, who would purchase his contract and be his master for a term of years. During that time the master would agree to feed, clothe, and house the servant and to provide him, at the expiration of the term, with seed, tools, and any other freedom dues mentioned in the indenture. Ordinarily when a ship reached Jamestown, servants were put ashore where assembled planters bought their indentures and took them away to their plantations; otherwise the servants were kept on board while the vessels sailed up the rivers to the plantations, and the owners came out to purchase them.[18]

Of the hundreds of people arriving in the colony annually, "scarce any but are brought in as merchandise to make sale of," Secretary Kemp reported from Jamestown in 1638. To work out one's passage was actually not an inhuman way for a poor man or woman to get to Virginia and learn the ropes before he was turned loose to fend for himself, or in the case of women, to marry or be employed in a household. From the point of view of the servants, everything depended on how the master treated them.[19]

London officials learned in 1623 "that divers in Virginia do much neglect and abuse their servants there with intolerable oppression and hard usage," and there was much truth in this report. Thomas Best wrote home to his brother that all members of the household in which he served were on the verge of starving: "My Master Atkins hath sold me for 150 £ sterling like a damnd slave as he is for using me so baselie." The widow Dickenson lost her husband Ralph in the "bloudy Masacre" when he had but three years left to serve his master, Nicholas Hyde; carried off by the "Cruel salvages," she endured great misery for ten months. For two pounds of beads provided by Dr. John Pott of Jamestown, she bought her release only to find that the physician linked her to servitude "with a towefold Chaine"—her husband's obligation of three years and the ten months of her imprisonment, the latter due for his furnishing her with the ransom—"tow much for two pounds of beads." Considering service for that well-known medical rogue no different from her slavery under the Indians, she petitioned the Council for release, but Dr. Pott was a member of that body and insisted that she serve "the uttermost day" unless she paid him 150 weight of tobacco.[20]

The "barbarous usuage" of servants by "cruell masters" grew to be "soe much scandall and infamy to the country in generall" that willing servants were hard to lure over the seas. Aware of this threat to the labor supply, the General Assembly sought to guarantee the servants adequate diet, clothing, and lodging, as well as to order that only "moderate correction" be administered by their masters and mistresses. Governor Lord Culpeper insisted in 1683 that it was "extremely difficult to keep an equal hand between bad masters and bad servants." The work that both black slaves and white servants were expected to do was "no other than what Overseers, the Freemen, and the Planters themselves" did. Generally, white servants received somewhat better food and clothing than the blacks, but the slaves were "not worked near so hard, nor so many Hours a Day" as the white farm workers. Undoubtedly the planters wanted to get all the work they could from servants whom they could keep but a few years—the term of bondage had been reduced to four years by the Assembly after 1660—whereas they were concerned about the health and endurance of their slaves.[21]

The incidence of mortality in Virginia throughout the seventeenth century fell heaviest upon the white laboring population. It has been estimated that indentured servants were landing in Virginia at the rate of 1500 per year after 1660. According to Governor Berkeley, in 1663 only one out of five newly arrived immigrants survived the first year; four years later Mr. Garroway declared that the colony was losing "a third part" of all newcomers. Of those fortunate enough to outlast not only the seasoning time but the following three years of hard labor and unsanitary living conditions, only one in ten, according to a leading authority, took up land when released from servitude; the rest succumbed to disease, returned to England, or became artisans or perhaps overseers.

Occasionally, fortune smiled upon an indentured servant, and we are lucky to have an odyssey written by just such a man who came to Jamestown sometime between 1656 and 1671. James Revel, a tinsman by trade, was sentenced by a London court to serve fourteen years in Virginia for the crime of theft. In the colony he was sold as a servant to a hard master living along the Rappahannock River. After "twelve long tedious years did pass away," the master died, and his widow put the plantation and its bound people up for sale:

> A lawyer rich who at James-Town did dwell
> Came down to view it and lik'd it very well
>   He bought Negroes who for life were slaves,
> But no transport Fellons would he have,
> So we were put like Sheep into a fold,
> There unto the best bidder to be sold.

This ballad, which is quoted for information rather than elegant versification, told curious Londoners that a compassionate craftsman bought James Revel and used him as a servant, not as a slave:

> My kind master did at James-Town dwell;
> By trade a Cooper, and liv'd very well:
> I was his servant unto him to attend,
> Thus God, unlook'd for raised me up a friend.

Thus did I live in plenty and at ease,
Having none but my master for to please,
And if at any time he did ride out,
I with him rode the country round about.
And in my heart I often cri'd to see,
So many transport felons there to be. . . .

When Revel's two years were up, the friendly cooper arranged for his passage back to England where, we assume,
he must have lived happily and gone straight ever after, and
earned enough money to pay hawkers for peddling his auto-
biographical ballad around Cheapside and the Fleet.[22]

Everyone who had anything to do with the village during
the seventeenth century accepted without question the hierarchical arrangement of all persons in English society. "The
better sort" consisted of the king, lords, and gentry; the
"ordinary sort of people" were the yeomen, husbandmen,
artisans, tradesmen, common laborers, and temporarily bound
servants. To the latter class, in Virginia, the alien and "inferior" body of black heathens was added.

From the first days of the settlement to the fall of the
Company in 1624, a predominantly male population lived under military or authoritarian government, and a normal English type of society could not develop. The most essential ingredient, the family, was almost nonexistent at Jamestown,
and where six out of seven men did not long survive, death
precluded any continuity of arrangements.

Jamestown did not develop into a metropolis for the abased
Caucasian many; rather it became the rendezvous of the fortunate white few for whose benefit the affairs and life of the
colony were ordered. Growing rich by the culture and sale
of tobacco, these few did not fall in the social scale; they rose

to the top. And as years passed, they achieved a privileged position and produced the first cohesive social class in the Old Dominion. A handful of this nascent gentry resided at Jamestown, where they presided over and worked for the special interests of their class in the midst of a motley floating population of artisans, publicans, mariners, servants, and slaves.[23]

Fondly remembered ways of Old England figured as soon as the opportunity permitted. The "adventurers and planters" in Virginia had stated in a petition to the Council at Jamestown in 1620 that "all cannot be select . . . Great Actions are carryed with best successe by such Comanders who have personall Aucthoritye and greatness answerable to the Action." Inasmuch as men of vulgar and servile nature cannot be swayed by "a meane man," they demanded leaders of "the better sorte," and a man of Quality for their governor. Here, certainly, is no testimony to democratic aspiration. Even the fact that George Yeardley had been knighted in 1618 before he was named Governor the second time did not commend this "meane fellow" to many of the leading colonists, who looked upon him as "the right worthie Statesman, for his own profit."[24]

Illustrative of the leading men of Virginia was George Menefie, who arrived there in the *Samuel* in 1622. It was only a short time after the "Massacre" and the settlers were busy adjusting to the changes brought about by the attack. In October 1629, Menefie was chosen a burgess for James City, and by 1635 this young "lawyer" (as he referred to himself) was a member of the Council of State and one of the leaders in deposing Sir John Harvey from the office of governor. In the "New Town" on the island, Menefie constructed his town house adjoining the properties of the merchants John Chew and Richard Stevens. By this time he was the town's principal merchant, trading in servants and tobacco, and, as mentioned above, he was probably the supplier of blacks for the first slave market at Jamestown.

As a trader in both servants and slaves, George Menefie was able to build up several landed estates by exchanging headrights for acres, one of them being "York Plantation," later part of Rosewell. On the James River not far below the town, Menefie acquired a fine plantation called Littleton, where De Vries was entertained by the great merchant in March 1631: "Here was a garden of one morgen [two acres], full of Provence roses, apple, pear, and cherry trees, the various fruits of Holland, with different kinds of sweet smelling herbs, such as rosemary, sage, marjoram, and thyme. Around the house were plenty of peach-trees, which were hardly in Blossom. I was astonished to see this kind of tree, which I had never seen before on this coast."[25]

Landed estates could be accumulated most easily after 1619 by means of the headright, which entitled anyone to fifty acres of land for each individual, whether freeman, servant, or slave, whom he brought into Virginia. Inasmuch as the warrants for land were issued by the secretary's office at Jamestown, masters of ships and merchants, chief mariners and others procured them easily on going ashore. The resident merchants at the port, like Master Menefie, were the ones most favorably situated for attaining the goal of most Englishmen in the colony, one or more plantations and the social prestige and political power that went with them.[26]

It astounds the modern person who reads the records to learn how quickly the colony became so unbalanced socially. Land, power, and privilege were the prerogatives of a very small group of individuals who, like George Menefie, had family and mercantile connections in England and ready access to political influence through the Governor and Council in Virginia. As early as March 1624, the General Assembly voted that delinquent "persons of quality . . . being not fit to undergoe corporal punishment" should be either fined or imprisoned, but not whipped.[27]

Sir William Berkeley frankly conceded in 1660 to their Lordships of the Council for Trade and Plantations the truth of "the imputation . . . that none but those of the meanest quality and corrupted lives go" to Virginia. "But this is not all truth, for Men of as good families as any Subjects in England have resided there . . . But wee will confesse that there is with us a great Scarcity of good men." Such good men as there were, however, were of two sorts: those whom Berkeley said in 1663 "keep looking back on England with hopes that the selling of what they have here will make them live plentifully there," and those who were content to remain colonists and to found what have been termed "the first families of Virginia." The genteel characteristics they displayed eventually were largely acquired in the Old Dominion; they were a folk risen not from the English gentry but *bourgeois plutocrats*, if you will, from a sound, aggressive, acquisitive middle- and lower-class stock. "Those mighty dons, those parvenues," Governor Francis Nicholson labeled them at the end of the century, ruled Virginia from Jamestown, and there they set the tone of its life. For the creating and maintaining of these tobacco planters and their families, a tremendous price was paid, not only in human lives of servants and black slaves but by the entire society as well.[28]

V

God and Man at Jamestown

The spirit of the times of the first two Stuart kings of England, and for the Roman Catholic Spaniards and Frenchmen as well, differs so radically from that of the twentieth century that it is almost incomprehensible to us for our attitudes and beliefs are worldly. They were the spiritual heirs of the Middle Ages. Everything that Englishmen did or wrote about colonizing was encased by religion and suffused with a genuine piety. Men might speak of national purpose or of pursuing trade and navigation, but they always professed to be seeking Christian goals. In every event they saw the hand of God, and their language was heavily Biblical. Such piety was not insincere; their missionary fervor was real.[1]

There were in England in the early decades of the seventeenth century many more puritans* than has been realized. These people, laymen as well as clerics, all believed that the Church of England still retained many Roman rites and cere-

---

* In this book, *puritan* is used to indicate an attitude toward God, the Church, morals, behavior, and a person so inclined; whereas *Puritan* refers to the members of a party or the radical wing of the Church of England.

monies that the Protestant reformers had failed to abolish. They divided over the issue of how to eliminate these "popish practices": the largest group desired to remain with the Church and "purify" its ritual and worship from within; the lesser number, despairing of internal reform, became "separatists" (like the Pilgrim fathers and, later on, the Puritans of Massachusetts Bay) and left the Anglican fold to found a new Jerusalem. The puritans who remained in the Church—conforming puritans—did not differ over theology or Christian fellowship with those who went out to New England to erect a Godly society; they differed merely over ways and means.

The men who established and managed the London Company were, to be sure, conforming Anglicans, but nothing is more evident in their writings than a strong puritan bent. The ministers whom they so carefully chose to go to Virginia were conforming clergymen, trained in the universities but nevertheless strong Calvinists, that is, puritans. Samuel Rawson Gardiner once observed that the first thing an Anglican parson of the early 1600s did on leaving England was to heave the Book of Common Prayer overboard. In 1610 the Company sent to Jamestown the Reverend Richard Buck, a puritan without doubt. It was he who officiated at the marriage of Pocahontas and John Rolfe in the church in 1614 and who served as chaplain when the first legislative assembly convened in the choir in 1619.[2]

Outstanding among the early Virginia clergy was the Reverend Alexander Whitaker, a cousin of a great London puritan, William Gouge, rector of St. Ann's Blackfriars. Master Whitaker came to Virginia in the same fleet as Master Buck, and his experiences in the colony did not in any way moderate his puritan views, for within a year he was informing a friend at St. John's, Cambridge: "We have no need either of ceremonies or bad livers; discretion and learning, zeal with knowledge would do much good."[3]

Because Whitaker served as minister at Henrico and was chaplain to Governor Dale, the latter called him to Jamestown to give Pocahontas Christian instruction and baptize her before her marriage to John Rolfe. The Indian princess thus acquired her Anglican faith in unadulterated puritan form. In a letter to William Gouge that same year, 1614, the young parson wrote: "I much more muse, that so few of our English ministers that were so hot against the surplice and subscription come hither where neither [are] spoken of." No better testimony can be offered to bolster the idea that in the long run the puritan outlook imported at Jamestown made low-churchmen of the Episcopalians in Virginia.[4]

Among the most important lay leaders of the colony were men of marked Calvinist temper. Captain John Smith, despite his gasconades, had qualities that commended him to the founders of New England. Foremost among the puritan rulers was Governor Dale, whose sabbath laws would have won approval from strictest English Puritans. Master Whitaker praised "our religious and valiant Governour," whose missionary zeal never flagged. Dale also sought, unsuccessfully, to marry one of Powhatan's daughters; it is no wonder that such devout Calvinists as Ralph Hamor and John Rolfe were strongly drawn to him.[5]

Undoubtedly it was John Rolfe who, above everybody else, displayed most of the virtues and lived the life of a complete puritan by his properly Christian attitude toward the natives, by working hard to improve the colony's economy and making tobacco the staple, and by long years spent in public service. Neither of those great New England men, John Cotton or John Winthrop, could have improved upon the spirit and diction of this puritan who dared to write to his King and the Company in 1616 about the hopeful state of Virginia: "What need have we then to fear, but to *go up at once* as a *peculiar people* marked and chosen by the *finger* of God to possess it? for *undoubtedly he is with us.*"[6]

Many of the chief men at Jamestown before 1624 would probably have been in accord with those at Plymouth and Boston had they met. Not so the rank and file of the James River settlement. Herein lay the profound difference between the two English societies in North America. William Gouge spoke for all the puritans when he described the *family* as "a little Church, and a little common-wealth." And Thomas Gataker taught that "this society is the first that ever was in the world." For all puritans, society was rooted in the family, and both church and state, being really one, were but the family writ large. Whatever the rulers at Jamestown may have thought, the temporary and unstable populace of this waterfront settlement, having few compelling religious convictions, tended to frustrate all piety and missionary endeavor. In New England, the Puritan cause succeeded because almost the entire population, ranged in families, favored it—even at the worldly port of Boston.[7]

Outstanding in their concern for the Indians were Master Whitaker and George Thorpe. It was their hope to teach them civility and to educate them that they might understand about God and learn the Gospel story. The natives could not grasp the idea of the Christian spirit; only Opechancanough with his Dominican and Jesuit training could comprehend it. Powhatan, Tomocomo, and the rest thought of God as a finite being. As was often the case, Christian missionaries betrayed a certain arrogance in telling a whole people that what they believed was all wrong. The heathens could, however, observe at Jamestown how far apart English ideals and English actions were. In 1622, when the Indians mounted their defensive war, George Thorpe was murdered, and the projected Indian school at Henrico was dropped.

In anticipation of the dissolution of the Company, the General Assembly voted in March 1624 that "there be an uniformity in our Church as neere as may be to the canons in England both in [substance] and circumstance." But this was not to be. The sad truth was that from a religious point of view, neither the King nor his officers of state, who governed this first royal colony after May 1624, nor the prelates of his Church attempted to promote "an uniformity" or to transfer the Anglican faith and institutions intact to Virginia. It may be said that the Crown abandoned the Church in the Old Dominion, leaving the laity and clergy there to organize and invigorate it themselves. In one respect this was a happy oversight, for the Virginians were able to work out all ecclesiastical problems in the light of their own experience in the colony. Because of this neglect by the Church of England, the settlers achieved religious self-government by default.[8]

The members of the Council and House of Burgesses proceeded between 1624 and 1643 to organize their church by passing a series of acts affecting ecclesiastical concerns.* The regulations for behavior on the Lord's Day made in 1629 resemble those of Puritan New England: people were not to profane the holy day either by working or "journeying from place to place," and all servants were required to attend divine worship. Encouragement was also given for the building and

* For the parish and vestry, see Chapter VIII.

repair of church edifices. It is apparent, however, that after the Virginia Company lost its charter, the strong puritan influence on life in Virginia began to wane.[9]

In 1630, when the Anglicans were attaining more prominence, the authorities ordered that all ministers had to conform to the Church of England. By this time the Council had become fearful of Lord Baltimore's plan to settle Roman Catholics in Virginia and reminded King Charles I in a petition that among their many blessings "there is none whereby it hath beene made more happy then in the freedome of our Religion, which wee have enjoyed, and that noe papists have bene suffered to settle their aboad amongst us." In 1641 an act was passed disabling "popish recusants" from holding office in the colony or any one who refused to take the oaths of allegiance and supremacy. Within fifteen years after the loss of the Company's charter, the Anglicans were actively persecuting not only Roman Catholics but Protestant dissenters of every stripe in their quest for religious unity. With the arrival in 1642 of Sir William Berkeley, an ardent Anglican and a strong governor, the tightening of church organization and the elimination of Puritans and others went on apace.[10]

Governor Berkeley and his Council began at once to harry all "seditious sectaries" and other "schismatics" out of the land. Between 1642 and 1648, Sir William and his "faction . . . which called itself 'the Church of England,' " according to Cotton Mather, persecuted them, summoning their ministers and others in Nansemond and Elizabeth City counties to appear before them at Jamestown. As a result, about a thousand of these "puritans" migrated to Maryland or New England. By 1650, when Sir William displayed his instructions from Whitehall to "Suffer no Invasion in matters of Religion," the toleration of diverse religious opinions within the church, which had been so conspicuous and acceptable, had given way under his aegis in the little capital to a narrow bigotry.[11]

In the year 1656, during the period of the Commonwealth in England, the advance missionaries of the new sect called the Society of Friends arrived in Virginia. Almost at once the old statutes directed at puritans and Roman Catholics were again put in force; and early in 1660 the Assembly passed an act for the suppression of an "unreasonable and turbulent sort of people, called Quakers." In large measure this law arose out of the case of William Robinson, who spent six months in jail at Jamestown for attempting to arouse Virginians "to be sensable of the power of God, and spirit of truth."[12]

Governor Berkeley back in office after the Restoration enforced this law and a later one of 1662 dealing specifically with the Quakers most vigorously. He dealt harshly with county sheriffs who appeared remiss in taking up members of "this most pestilent Sect" and for not sending the most refractory prisoners to Jamestown. Half of the heavy fines collected from violators was allotted to informers. Two Friends, William Cole and George Wilson, came to Virginia while this campaign to drive the Quakers out of the colony was going on, and they were arrested and put in the "loathsome prison" at Jamestown. Reporting his experience to the Quakers in New England in September 1661, George Wilson told them that he had been "chained to an Indian which is in prison for murder; we had our legs in one bolt and made fast to a post with an ox chain, but not now, though [still] in irons."[13]

The second letter sent "From that dirty dungeon in Jamestown, the 17th of the Third Month 1662," is more specific in describing the indignities and horrors which prisoners had to endure, but it deserves to be quoted: "If they who visit not such in prison (as Christ speaks of) shall be punished with everlasting destruction, O what will ye do? Or what will become of you who put us into such nasty, stinking prisons, as this dirty dungeon, where we have not had the benefit to do what nature required, nor so much as air, to blow in, at a window, but close made up with brick and lime, so that there is no air to take away the smell of our dung and piss, who for all their cruelty I can truly say, 'Father forgive them for they know not what they do.' But thus saith the Lord unto me: 'Tell them that because wilfully they are ignorant. I will strike them with astonishment, and will bring upon them the filth of their detestable things, and in that day they should be glad, if they could, to eat their own dung and drink their own piss, it shall so odiously stand before them, that it shall be an evil stink in succeeding generations. This you shall eternally witness . . . this is to go abroad.' "[14]

If Wilson's denunciation seems crude and too frank, the reader must be reminded that what he is telling us, unpleasant though it be, is a part of American history. In *The Sufferings of the People Called Quakers* (1703), Joseph Besse reveals that George Wilson died in the Jamestown prison "in cruel irons which rotted his flesh."[15]

Many other Friends were incarcerated at Jamestown and presumably underwent similar painful experiences. The great Quaker preacher, William Edmundson, visited Governor Berkeley at Jamestown in 1671 to discuss "Friends' suffering"; he told him how kind his brother John, Lord Lieutenant of Ireland, was to Quakers in the hope that he would be persuaded to treat the ones in Virginia charitably. "He was very peevish and brittle; and I could fasten Nothing upon him with all soft Arguments I could use; so . . . I left him."[16]

As the principal community and seat of authority for the colony, Jamestown was also its religious center. In 1607 after constructing some fortifications and arranging for temporary shelters, the primary concern of the leaders was to provide a place in which to hold divine services. In the first busy weeks, they used "a rotten old tent" until they could put up a barn-like structure. It lasted less than a year, burning early in 1608. When Governor Argall surveyed the post in 1617 he found that the second church, erected in 1608 and repaired in 1611 had fallen down and the colonists were using a storehouse for their services. He promptly supervised the building of a third edifice, one of timber construction measuring 50 feet by 20 feet that rested upon a foundation of brick and cobble.[17]

No building lasted very long in Jamestown, and so, under Governor Harvey, a fourth church, to be made of brick, was started and supposed to have been completed in 1647. Described in 1676 as "faire and large," this structure was the one Nathaniel Bacon personally set fire to later that year; its woodwork was burned and possibly the entire building. During the next ten years, at the expense of the parish of James City County, a fifth church (55 feet by 28 feet) rose—possibly a restoration or rebuilding—which was used until the conflagration of 1698. Whatever criticism may be leveled at the faulty construction and design of buildings at Jamestown, in this structure the capital could boast of having the best church edifice in the colony.[18]

To make known to the mother country the unhappy state of the church in Virginia the Reverend Roger Green published *Virginia's Cure* in 1662. He emphasized "the great want of Christian Neighborhood, or brotherly admonition, of holy Examples of religious Persons, of the comfort of theirs, and their Ministers' Administrations in Sickness and Distresses, and of the Benefit of Christian and Civil Conference and

Commerce." Such deficiencies could be largely attributed to the dispersed nature of plantation settlements, but at James-town and the mainland environs of its parish, the lack of any numerous permanent body of inhabitants seriously affected the well-being of the church.[19]

Throughout the century, the quality of available ministers was never very high, as numerous contemporaries pointed out. Sir William Berkeley, a devout churchman, declared to the Lords of Trade in 1671 that Virginia parsons were well paid and that he would approve increasing their remuneration "if they would pray oftener and preach less. But of all other commodities, so of this, the worst are sent us, and we had few that we could boast of." Lord Culpeper echoed these senti-ments to their Lordships when he was Governor in 1683: "Good ministers would in time certainly gain you a hold on the people, but there is nothing to encourage good men to go so far."[20]

Master John Clayton was, perhaps, the ablest man who ever served at the Jamestown church, and we are fortunate to have some information about him. He was a graduate of Oxford where, at Merton College, he acquired an enthusiasm for natural science and medicine. About 1682 he sailed in the *Judith* for Jamestown to be the incumbent of the court church, the best curacy in the colony.[21]

His first letter, written on April 24, 1684, after he had landed, was to Nehemiah Grew of the Royal Society of Lon-don in which he described his reception: "Tis now our Great Assembly and on Sunday, by a peculiar order from the Gove-ner and Councell, I am to preach, so that something peculiar is expected, and I must mind my hits to preserve that blooming repute I have got. I have had the happinesse to be cried up farr beyond my deserts. The people are peculiarly obligeing, quick, and subtile." That the sermon pleased the officials is evident by the £5 sterling allowed him by the House of Burgesses. In a more serious vein, he gave his impression of the colony: "in short its a place where plenty makes poverty, Ignorance, ingenuity, and covetousness causes hospitality."[22]

Master Clayton had apparently accepted the charge at Jamestown because it offered the opportunities to pursue the scientific inventory of Virginia requested by his friends at the Royal Society; but unfortunately, all of his books and instruments were lost at sea. His parish work took up very little of his time, so he was able to devote himself to studying the country, its diseases, and its people.

Clayton lived on a tobacco and stock-raising plantation, owned by a widow, that was located close by the marshes of Jamestown Island. Much interested in farming, the parson tried in vain to induce the overseer of the plantation to adopt better methods of cultivating tobacco and to drain the marshes to create additional arable land. Getting no response to his suggestions, he took a few of the plantation's servants and drained some of the swamp; then, using his own method, planted tobacco that produced five times the seedlings yielded by the overseer's methods. He also advised keeping the island's cattle in barns in the winter months and advanced a plan for building a better fort to protect Jamestown.

Most of what he learned is to be found in letters and papers he sent to the Royal Society in after years. In one he described accurately a prevailing malady that he named the "Colick," but which the colonists called the West Indian Dry-Gripes. What he did not know was that it was actually lead poisoning caused by drinking rum that had been distilled in lead pipes. When the Royal Society published Clayton's contributions in *The Philosophical Transactions* for 1739, they provided excellent descriptions of the Old Dominion. In return, posterity can perhaps forgive his neglect of the cure of souls.[23]

The inability to procure competent and dedicated clergymen in sufficient numbers, as well as the solving of every other ecclesiastical problem of the Virginians, was the consequence of the inefficacy of the crown and the Anglican prelates. Presumably the colony was under the jurisdiction of the bishop of London. After the Restoration, prominent Virginians, recognizing the need for more authority over the clergy and laity than the vague supervision they were getting, devised some proposals.

The deputy governor in 1662, Francis Moryson, asked the Assembly to "set downe certaine rules to be observed in the government of the church, until God shall please to turn his majestie's pious thoughts toward us, and provide a better supply of ministers." At this same time, the Reverend Roger Green urged that a bishop be sent over "so soon as there shall be a City for his See." Four years later Secretary Thomas Ludwell in "A Description of the Government of Virginia," composed for royal officials, said frankly: "I could hartely wish that my Lord of London, and other greate Clergymen, would take us a little more into their care." He conceded that conditions at Jamestown were still "unfitt for a Bishop to reside" there, and because "we are subjected to the See of London," there are no superior clergy in the colony.[24]

Complaints such as these from important colonists made it possible for some persons of great influence at court to win the support of the lord chancellor, the Earl of Clarendon. A draft of a royal proclamation, October 16, 1673, probably framed by Sir Leoline Jenkins, placed Virginia for the first time in an established position within the Church of England

and under the archbishop of Canterbury. It also made provision for the creation of a Diocese of Virginia and, significantly, annexed to it all the "remaining regions and plantations (except New England) in America" and the West Indies. The proclamation stated further that "the place of James City and of the church there be created . . . as an Episcopal See and a Cathedral Church . . . and that all our city, Jamestown, be from now and henceforth forever a city . . . named the City of Jamestown." The first bishop was to be Dr. Alexander Murray (Moray), a Scottish clergyman and devoted follower of King Charles in 1652.[25]

Although approved by the King-in-Council in 1673 with letters patent for the execution of it, the proclamation was never issued. Had it been, the presence of the bishop for most of the English colonies in America would have certainly improved ecclesiastical conditions in Virginia and conferred an enormous prestige on its capital, which in fact would thenceforth have been correctly called the "City of Jamestown."[26]

Meanwhile the colonial church continued under the bishop of London who, except for licensing ministers to go to Virginia, had virtually no other authority. When Henry Compton became bishop in 1675, he realized the pressing need to remedy the many moral and ecclesiastical abuses in the plantations and to regularize the churches according to the canons of the Church of England. Bishop Compton decided to delegate such authority as he had over the colonies to commissaries who should reside in the colony. In time, he appointed "James Blair, Clerk," then rector of the remote parish of Henrico, to be commissary of Virginia, who assumed his duties in June 1690.

Master Blair, probably with the advice of Francis Nicholson, the new governor, immediately called a convention of the Virginia clergy at Jamestown. It met on July 23, 1690, and dealt with two matters deemed essential to the welfare of the church in Virginia. To ensure a supply of well-trained native-born clergymen, plans were made for the founding of a "seminary of ministers of the Gospel," which, in 1693, was chartered by the King as The College of William and Mary and situated at Middle Plantation (Williamsburg) in Bruton Parish. The other decision was that the commissary should issue "an order for the more convenient execution of ecclesiastical discipline."

The commissary and the convention, by these actions, interpreted the authority conferred on James Blair as allowing him to do in Virginia everything a bishop of London could do except ordain priests and confirm individuals. Master Blair, however, overstepped the bounds when he proclaimed his intention to establish ecclesiastical courts for dealing with the morals and behavior of both clergy and laity. Jamestown, with its floating population, was to have one of these ecclesiastical tribunals, but the plan drew a severe rebuke from the House of Burgesses in May 1691, for such courts were anathema to low-church Anglicans.[27]

A second convention of the clergy met in Jamestown on July 25, 1696, to deal, principally, with the refusal of the Assembly to increase ministerial salaries and perquisites. The burgesses denied the proposal, for, as they said, clergymen in the colonies "are in as good a condition in point of livelihood as a gentleman that is well seated and has 12 or 14 servants." Commissary Blair and the other delegates denied that the members of the convention were as well off as the burgesses so "ornamentally" described. This was a question of religious and ecclesiastical importance at Jamestown in the months before fire destroyed the village in 1698.[28]

James Blair favored Middle Plantation and Bruton Parish over Jamestown on all occasions, and by procuring the location of the college at the former place, he shares with Governor Nicholson the credit or the obloquy for the religious, as well as the political, eclipse of Jamestown.

VI

Self-government and Self-interest
in a Planters' Parliament

In our early schooldays, most of us were told that "the first legislative assembly in America" met at Jamestown in Virginia in 1619. Edward Channing wrote of this body in his celebrated *History of the United States* that "it was, indeed, the 'mother' of the American representative legislature." This may seem strange at first, for there was nothing *democratic* about English society in that age, and one does not expect to find it in the colony of Virginia. Democracy was not even an aspiration then, for most men agreed with John Cotton that it was not "a fit government either for church or commonwealth." Therefore until we know exactly what took place in Jamestown before 1698 and how this first legislature developed there as a political institution, then and only then can we attempt to assess its symbolic importance for later American history.[1]

During 1618 the Virginia Company worked out a program of reform for the management of its colony, which has been referred to as "The Great Charter." Previously the settlers had been living under authoritarian rule, and most of them were, or had been, servants of the Company without any private property or legal rights. Many of them, moreover, harbored numerous grievances. To reduce such complaints, win the support of the planters, improve the reputation of the colony in Britain, and stimulate migration to Virginia, Sir Edwin Sandys and his associates planned radical changes in the colony's government.

The account of their deliberations and their decisions was set forth in "A Breif Declaration of the Plantation of Virginia," a document approved by the general assembly in 1623 at Jamestown. It reveals the kind of life the "ancient planters" led and what the survivors thought about it. The authors of the "Breif Declaration" emphasized that Governor Sir George Yeardley, soon after his arrival at Jamestown in April 1619, issued a proclamation freeing all persons who had resided in Virginia prior to the departure of Governor Sir Thomas Dale in 1616. He also released them from forced public service and labor, which "formerly they suffered, and that those cruell lawes by which we had soe longe been governed were now abrogated, and we are now to be governed by those free lawes which his Majesty's subjects live under in England." Martial law was replaced by English common law.[2]

A more salient point, which promised much for the future, was that in order that they "might have a hand in the governing of themselves; it was granted that a general assembly should be held yearly once, whereat were to be present the governor and council and two burgesses from each plantation freely to be elected by the inhabitants thereof." The assembly would have the power to make "whatsoever lawes and orders should by them be thought good and profittable" for the subsistence of the colony.[3]

Governor Yeardley promptly called for an election, and such meager evidence as we have suggests a fairly free franchise. In every instance, as one might expect, the burgesses chosen were the leading planters, probably by acclamation in an open forum. Seven of the thirty-one men who attended the first meeting lived in Jamestown or the immediate vicinity and exemplify the composition of the body. The burgesses elected to represent the town were Captain William Powell and Ensign William Spense who had been in command there under the military rule and who continued to govern the post in that respect for several years. John Pory had lived in the village but a short time but was secretary of the Council; Thomas Pierce acted as sergeant-at-arms for the Assembly, and the Reverend Richard Buck served as chaplain. The Governor, Sir George Yeardley, was present, as was John Rolfe, both of them members of the Council and both residents of Jamestown.[4]

The first session of the General Assembly was held from July 30 to August 4, 1619—"the most convenient place we could find to sit in was the choir of the church," John Pory informs us. The Governor took his accustomed place with the councilors on either side of him. Secretary Pory, having been appointed speaker, sat in front of the Governor with John Twine, the clerk, on one side and Thomas Pierce, the sergeant, on the other. While Master Buck offered a prayer, the burgesses stood in the choir; afterward they took regular places in the body of the church. Like all Englishmen, these dignitaries cherished ceremony, and, guided by John Pory who had been a member of the House of Commons, they followed the procedures of that body as closely as they could. In one respect, however, they copied the court of the Company, whose treasurer, councilors, and adventurers sat as one body; not until 1663 did they divide and sit as two houses.[5]

Convening in Jamestown in the last days of July, which in

Virginia are generally hot and humid and enervating, the newly formed Assembly settled down to discussions of pressing local issues. The first order of business was to examine the credentials of each burgess, thereby setting the precedent for succeeding American legislative bodies to be the sole judges of the qualifications of their own members. Next came the matter of relations with the Indians, which had worsened. The Assembly issued a warning that "no injury or oppression be wrought" against the natives. Ensuring a supply of European goods for the colony, tobacco problems, taxes, and a number of private acts were all tackled in this first week of the first session of the first legislative assembly in America.[6]

The governor, even though he sat in the Assembly, had the right of veto over every act passed by it; furthermore, no act could go into effect until it was approved by the Company in London. In the authorization for a General Assembly, however, was a memorable provision stating that once the colonial government should be well established, "No orders of our Court [the Company's] afterwarde shall binde [the] colony unles they bee ratified in like manner in the generall Assembly." Here was a promise of self-government. Although it was repeated in the royal instructions sent to Governor Sir Francis Wyatt in 1625, the Company did not last to fulfill the promise.[7]

Throughout the century, the political leaders of Virginia persisted in making self-government their goal, and the Company's promise provided the opening for many future assertions of the assembly's full competence. As Herbert L. Osgood cogently observed in 1904: "The transition of a community within a decade from a state of subjection, such as that portrayed in the writings of [Captain John] Smith, or in the *Lawes Divine, Morall and Martiall*, to a condition such as that suggested by the promise of the Company would awaken surprise in any age of the world, and most of all, perhaps, in the seventeenth century."[8]

For more than ten years after the fall of the Company in 1624, the fate of the General Assembly remained in doubt. In 1625 the Virginians petitioned the new King, Charles I, to be allowed to retain their legislature, but without success. Several times royal governors called sessions to pass laws for special purposes, but the institution had no permanent legal status before 1638 when the King instructed the Governor, Sir Francis Wyatt (who had been appointed for a second time) to call an Assembly every year. This decision established the principle that every royal colony created thereafter should have an elected Assembly, not of right but "by royal grace and favor." Through insistence and persistence, the Virginians had won a victory: a trading company had been transformed into a permanent liberal institution by a monarch known for his arbitrary rule at the time when despotism rather than freedom prevailed in Western Europe.[9]

The assurance of an annual meeting of the Assembly was indeed remarkable considering the times, but nothing could be farther from historical truth than the assumption that the Assembly of Virginia was either *democratic* or *popular* in the manner of our national and state legislatures of today. In fact we may even properly inquire how *representative* the General Assembly really was, not only in 1619 and 1639 but in the decades that followed. Certainly as long as Jamestown was the seat of the government of Virginia, the members of the Assembly, by their acts, used it to provide and preserve the legal foundations, as well as the political power, of the colonial plutocracy of tobacco planters and merchants.

A prime issue to be decided, one which the Assembly declared to be its prerogative, was the matter of taxes needed to meet the expenses of government. Notwithstanding the fact that the Governor could veto any act initiated and passed by it, the General Assembly made clear to a succession of royal governors between 1624 and 1643 that it alone had the sole right to levy taxes. In the revision of the statutes made early in the first term of Sir William Berkeley, the legislature "enacted and confirmed, that the Governor and Council shall not lay any taxes, impositions upon this colony, their lands or commodities otherwise then by the authority of the Grand Assembly."[10]

The members of the Assembly generally shifted the burden of public charges onto the people as a whole. At its first session in 1619, a small head tax had been voted to pay the expenses of the principal officers; and in 1629 the Assembly passed an act for a tax of "five pounds of tobacco per poll levied throughout the colony." By such means the incidence of taxation fell on the "poorer sort" of people, while each of the small clique of planters, who possessed acres and acres of land (the chief form of wealth), paid no more than the smallest farmer.[11]

Not surprisingly this equal tax on unequals aroused resentment. After the beginning of annual sessions of the Assembly, the outcry against the poll tax was such that in 1645 the Assembly was told in no uncertain terms that "the ancient and usual taxing of all people in this colony by poll equally has been found inconvenient and is become insupportable for the poorer sort to beare." The poll tax was then repealed and levies on property and tithable servants substituted. The planter oligarchy had been caught off guard, and not until October 1648 could it muster sufficient power to repeal the property tax and levy taxes only "by the poll." Thereupon the building of great estates went on apace, and broad acres were never taxed; and because the Assembly had established all of the colony's legal institutions (with the idea of furthering and protecting the concerns of the tobacco plutocracy), this body can be most accurately described as *a planters' parliament*.[12]

Similarly those tobacco magnates who, by means of power, social prestige, and riches, gained a seat on the Council regarded it as a source of political privilege. After 1640 the councilors not only saw to it that they themselves were exempted from all public charges (except church duties) but no levy by polls was laid on the first ten of their tithable servants. In this and other ways a nascent gentry ensured its power, property, and privilege.[13]

The establishing of a new social order in Virginia was no small undertaking; it took a long time to accomplish. The consolidating of political power in the hands of the tiny minority of rich tobacco planters was well along by 1640. They were forming a ruling class, but they still exhibited the haughtiness—some said arrogance—of self-made men. Such presumptuousness was bound to be revealed in relations with the mother country, for they did not yet display the *noblesse oblige* of a true aristocracy whose members were sure of their position at the top of society. Commenting upon the failure of the Assembly to accede to a request by the King to limit the production of tobacco in 1638, Governor Harvey and his Council scored the burgesses: "Wee still find you in love and league with your owne humours and in a resolute crossness to his Majesty's propositions."[14]

The institutionalizing of the merchant-planter oligarchy took place in Jamestown. The Governor, Council, and burgesses, the apex of the political system of Virginia, met there together until 1663 as the General Assembly. Afterward they separated, with the burgesses sitting apart from the Governor and Council. At the provincial capital a visitor also found the general court, consisting of the Governor and Council; and there also met the court of James City County. At the sessions of these tribunals, especially those of the general court, "men of the greatest abilities both for judgment and integrity doe usually meet." Their common interest in political, judicial, and business concerns was an important ingredient in uniting them into a recognizable social class.[15]

Every day, or so it seemed, the village was the scene of Virginia's bustling (and occasionally questionable) political and administrative activities. The secretary of the colony was the most important single permanent official, and he and his clerks lived in Jamestown and had their offices there. It was the secretary who issued the fifty acres of land for each head-right, and therefore he was constantly being importuned by merchants, ships' captains, and planters who had brought individuals into Virginia and were claiming parcels of land. Consequently the incumbent profited immensely from fees and other perquisites, and the office became a political prize; the position of sheriff of James City County was another lucrative post, the fees and other income to be derived from operating the country and county prisons attracting office seekers.

Sir William Berkeley, whom King Charles I commissioned Governor of Virginia in 1641, answered perfectly the description of the petitioners of 1620 who said they wanted a man of quality for Governor. Master of arts from the University of Oxford, reader of law at the Inns of Court, gentleman of the privy chamber of the King, author of a tragedy, "The Lost Lady," and brother of John, first Baron Berkeley of Stratton, he added a much needed luster to Virginia. His leadership, social and political, of the great planters was what was called for to complete the formation of a ruling class, even though he was unable to create a sense of community among all of the colonists.

During his first administration, 1642-52, Sir William enjoyed great success, chiefly by immediately identifying himself with the colony and energetically advancing its interests. Nothing attests better to his political acumen than the firmness he displayed in his first two years in restraining greedy plutocrats in their attempts to acquire excessive land grants. Then and also in his second term (1660-77), he proved to be both an able executive and an honest man. At first from Jamestown and later from his mansion at Green Spring plantation nearby, he handled such problems as controversies with the Indians, difficulties in the economy, and relations with the Crown and Commonwealth with understanding and decisiveness.

After the Restoration in 1660, following the end of the English Civil War, the Assembly, remembering him as a Virginia patriot, re-elected Berkeley whole-heartedly, and King Charles II, because of Sir William's loyalty to him during the trying years of the Commonwealth and Protectorate, approved his appointment to a second term as governor of Virginia. The new Governor saw to it that some of his loyalist friends, who had crossed to Virginia after 1652, had seats on the Council, and he appointed one of them, Philip Ludwell, an able though unscrupulous partisan, as secretary of the colony and the recipient of all the emoluments of the office.[16]

The Virginia Assembly elected in 1661 contained a majority of planters strongly attached to Berkeley's economic and political program to revive the colony's life. Burgesses were chosen by the freemen of their county whose sole "privilege" was to pick their legislators from the plutocratic few.

New members had often served on parish vestries or county courts, and their training in those bodies had strengthened their convictions of the need to maintain their class. Whatever had been their ideas of responsibility to the entire electorate, once they entered the State House on the island, the burgesses gave more thought to the affairs and desires of "the better sort" than to those of the increasing population of small yeomen farmers, indentured servants, and slaves. There was no change in this attitude on the part of the House of Burgesses under Governor Berkeley, for fifteen years passed before another election.

During this time, the long-standing complaints about the injustice of the poll tax built up and were studiously ignored by the privileged assembly. Francis Moryson, the colony's agent in London, told the Earl of Clarendon in 1663 that "all taxes (my Lord) with us are by Pole, not Acre," with the poorest and the richest "all paying equall." Another Virginian informed royal officials that "a poor man, who had only his labour to maintain himself and family, paid as much as a man who had 20,000 acres." In the taverns of Jamestown, much frequented by the tobacco magnates who favored the poll tax, there was general agreement with the special pleading of the Governor and Council addressed to the King in 1667 that he consider the Virginians (meaning the planting gentry) "as a people pressed at our backes with Indians, in our Bowells with our servants and poverty (brought on us by the hard dealing of those whome we are bound to defend), and invaded from without by the Dutch."[17]

The unabashed yearning for unrestricted self-government by the small privileged class was always more apparent at Jamestown and Green Spring than about the countryside. The General Assembly had grown into a planters' parliament, and to protect itself from outside pressures and ensure freedom of action in all of its affairs, the tobacco oligarchy, led by Governor Berkeley and the House of Burgesses, dispatched three men as agents to London in September 1674. Their object was to obtain a charter from the Crown that would define the rights of the colonists in the Old Dominion. During the distractions of the Civil War and Interregnum in England, the leading planters of Virginia had enjoyed virtual independence. Now they wanted the authority of the General Assembly sanctioned, to be assured of the "rights of Englishmen," and a guarantee against taxation from outside without the consent of the Assembly, and also royal confirmation of their property. What Thomas Ludwell, Francis Moryson, and Robert Smith sought at Whitehall was "to incorporate Virginia in the name of the governor, council, and burgesses alone." In this spirit of '74, the Virginia gentry were contending for a status expressed a century later in "the spirit of '76."[18]

The agents' task was complicated because at this very time the officers of the Crown were moving to tighten England's control over the colonies rather than to grant more freedom—even in the Old Dominion. King Charles II twice approved a charter as requested and then changed his mind, and the Privy Council ordered the instrument annulled. On October 10, 1676, he did grant a charter for the colony that provided for the dependence of Virginia upon the Crown and conspicuously omitted any acknowledgment of the authority of the Assembly, which continued to exist only by royal grace and favor. Evidently the three agents had confined themselves to representing the Governor and Council only, ignoring the burgesses.[19]

The failure to obtain a charter satisfactory to all three branches of the government plunged the men of Jamestown into deep gloom. Before they could conceive any way to achieve their desired goal, Bacon's Rebellion* took place, which, though suppressed by Governor Berkeley, ended all hopes of reviving the project—Berkeley was recalled to England and died soon afterward. The Privy Council made clear to other royal officials its own view, which was that "the Rebellion of Virginia was occasioned by the Excessive power of the Assembly."[20]

From 1677 to 1698, royal authority was quietly but steadily extended in the colonies, making these years very difficult for members of the Virginia Assembly who were determined to achieve legislative independence. The governors who were sent over endeavored, with considerable success, to discipline the "rebellious assembly" and expand the royal prerogative. Lord Howard of Effingham reached Jamestown in February 1684 and waged an unrelenting struggle for four years against the pretensions of the Assembly. In self-defense the House of Burgesses dispatched its champion, Robert Beverley, to London with a petition for the recall of his Lordship from office—and also from Virginia. Before King James received this plea, the "Glorious Revolution" swept him out of office, and it also marked a turning point in the history of the provincial legislative assemblies of all the colonies: thereafter their permanence was assured; their right to initiate laws became general; and, in imitation of the English Parliament after 1695, all were bicameral bodies.[21]

We have seen that throughout the seventeenth century at Jamestown, the Assembly, the existence of a fairly broad franchise in the colony notwithstanding, was never truly representative nor was it in any way a popular body or demo-

---

* See Chapter VII: Bacon's Uprising, Lawrence's Rebellion, 1676.

cratic. It was a planters', not a people's, parliament, a forum where tobacco magnates might shine before their peers as all the members looked after the well-being of their class. "There were no lords," a Huguenot visitor reported in 1698, "but each is sovereign on his own plantation. The gentlemen called Cavaliers are greatly esteemed and respected, and are very courteous and honorable. *They hold most of the offices in the country* . . . They sit in judgment with girded swords."[22]

The General Assembly had been introduced by the London Company in 1619 as a device to gain local support in Virginia for its new policies. Gradually it was transformed into the legislative Assembly granted by the King to all succeeding royal colonies. In the long perspective of history, it was the Virginians' finest political and legal accomplishment of the century. In after years, the institution was broadened and transformed to represent the needs and aspirations of all the people—self-government for all the people—many of them selfish, many of them not, the democracy of today. The legislative Assembly was to become a regular feature of government in the United States, and with this institution, English parliamentary ideas and practices developed in America.

VII

Bacon's Uprising, Lawrence's Rebellion

An insurrection of Virginia frontiersmen was transformed into the first popular revolt in American history on June 6, 1676, in a public house at Jamestown. What is known today as Bacon's Rebellion was too complex to be told here in its entirety, and in the pages to follow only what happened in the colony's capital and its immediate environs will be treated in detail. The revolt has been far too long polarized around "the governor and the rebel," but by concentrating attention on Jamestown, it may be possible to place both the issues and the leaders in a new perspective.

Later, as Virginians looked back on this time, many of them recalled that about 1675 there had been three "Prodigies" that were "Ominous Presages" of the disasters that soon befell the colony: a very large comet appeared in the sky every evening for a week; flights of pigeons so numerous as to break the limbs of great trees when the birds alighted for the night; and "Swarms of Flyes about an Inch long and as big as the Top of a Man's little finger rising out of Spigot Holes in the Earth." Belief in portents was universal in this age, but the truth was that the troubles that plagued Virginians were all of them man-made.[1]

Most historians agree that the troubles started on the Virginia frontier. Beginning in the summer of 1674, the settlers there clamored in vain for some decisive action by Governor Sir William Berkeley to curb the nearby "friendly Indians" and particularly the more unfriendly distant tribes who had killed above two hundred white men and harassed many others before May 1676. Sir William's dilatory and irresolute policy toward the hostile natives was resented by the frontier folk, who attributed "the Misteryes of these Delays" in part to the participation of the Governor and his favorites in the profitable Indian trade.[2]

Despairing of any prompt official measures, the people "rose in their own defense." In April 1676 a gathering of settlers in Charles City County persuaded an impetuous, youthful member of the Council of State, Nathaniel Bacon, Jr., to lead them against the Indian enemy. The young man was an English gentleman, sometime student at Cambridge and a newcomer to the colony (1674), whom Sir William Berkeley had promptly admitted to the Council. Much of his knowledge of men and conditions in the colony Bacon seems to have acquired on his visits to Jamestown, where he lodged at the large public house on Back Street, near the State House, maintained by the rich wife of Richard Lawrence. To this inn, wrote Thomas Mathew, an acute observer and a former lodger, "resorted those of the best quality, and such others as Businesse Called to that Towne." Lawrence, a man of "even Temper," was an Oxonian and a gentleman, landholder, and merchant who "made his Converse Coveted by Persons of all Ranks so that, being Subtile, and having those advantages he might with less Difficulty discover mens Inclinations and Instill his Notions where he found these wou'd be imbib'd with greatest Satisfaction."[3]

Early in 1676 Bacon's second plantation at the falls of the James was attacked by the Indians, and they killed his "much Loved" overseer and a servant. Before this raid, the planter had been content to be a mere spectator of events—showed little interest in politics while dwelling in the back country and had attended only two meetings of the Council.[4]

The pressure of friends among the gentry in the neighborhood, added to his anger over the killing of his overseer, induced Nathaniel Bacon to accept command of the men of the back country in April 1676. They were determined to proceed against the natives with or without a commission from the Governor, and when one was not forthcoming, Bacon led his forces into the forest and before long slaughtered many of the "friendly" Occaneechee tribe. For his defiance, Sir William Berkeley, on May 10, declared Bacon to be a rebel but offered free pardon to all of those who followed him if they would lay down their arms. At the same time he moved to improve his own authority by ordering an election of burgesses (the second in fifteen years) to whom people might present complaints against his government.[5]

The new "Grand Assembly" met at Jamestown on June 5, 1676. The next day, accompanied by forty-odd armed men, "Nathaniel Bacon the rebel" came down the James River in a sloop to represent Henrico. When he asked permission to land and take his seat, his party was fired on. That night Bacon slipped ashore in the dark and went to Lawrence's hostelry where he, his host, and William Drummond conferred in secret for three hours. In after years the non-partisan Thomas Mathew, who was in town as a burgess from Stafford, wrote of Drummond that he was "a gentleman of good repute," a sober Scot whose "wisdom and honisty . . . contending for superiority" were such that one could not judge him by ordinary standards. He had served as the governor of the Albemarle settlements from 1664 to 1667, had differed with Berkeley over land grants in Carolina, but he had long been a resident of Jamestown. John Culpeper, a noted questioner of authority in the Carolinas had come from Charleston in May, presumably to meet with Mr. Drummond about parlous affairs in Albemarle, and it is just possible that he too might have been present at the meeting.[6]

These men were among the most cultivated, best informed, and influential figures in politics, as well as in trade, of the planting gentry of Virginia. Although details are lacking, from what happened afterward we may reasonably surmise what was discussed and planned during that midnight session in a smoke-filled room: the Indian situation was undoubtedly canvassed first, and surely they rehearsed the many grievances, not merely of the inhabitants along the frontier but of all Virginians, themselves included, which had been mounting for at least a decade and a half. Evidently they persuaded Nathaniel Bacon to broaden his vigilante movement into an all-inclusive and active protest against the arbitrary government of Sir William Berkeley, which three times had nearly brought about conflict in 1673-74. By this maneuver, the gentlemen planned to expand a defiance of authority by backwoodsmen into a general revolt calling into question the entire government of the colony—provincial, county, parochial.[7]

These men were not seeking social change or anything resembling democracy; they were conspiring to bring together by the action of the new assembly a variety of causes under the aegis of political reform—*one great popular cause* that would appeal to Virginians of all classes: an insurrection against the dictatorial, incompetent, corrupt, and unprincipled administration (as they saw it) of Governor Berkeley and his small clique of councilors and officials, many of whom lacked birth and breeding. Despite his eloquence and élan, Nathaniel Bacon was "too young, *too much a Stranger there*, and of a Disposition too precipitate" to manage things to the degree that the other gentlemen planned "had not thoughtfull Mr. Laurence been at the Bottom," Thomas Mathew wrote after the events; but Bacon's sword, his armed men, overwhelming popular support, and the new "Grand Assembly" would be the conspirators' political engines. Such a combination, they thought, would be unbeatable.[8]

As early as April 28, 1676, Giles Bland, a former collector of customs whom the Governor had dismissed for quarreling with his favorite Thomas Ludwell, perceived that a genuine rebellion might develop out of the troubles with the Indians. This time, he declared, the freemen "are headed and Ledd by *persons of quality* there, which was wanting to them in 1674 when they were suppressed by a Proclamation, and the advice of some discreet persons that had then an Influence upon them; which is now much otherwise; for they are at this time Conducted by Mr. Nathaniel Bacon, lately sworne one of the Councell, and *many other Gentlemen of good Condition*, soe that it may be fear'd that the Enemie [potential rebels] will make a great advantage of these Disorders in the Government . . . ."[9]

Although the "other Gentlemen" took no overt action immediately, Thomas Mathew, who was present at the meetings of the new Assembly, stated unequivocally in "The Beginning, Progress, and Conclusion of Bacon's Rebellion" that "the received Opinion in Virginia . . . very much Attributed the promoting [of] these Perturbacions to Mr. Laurence; and Mr. Bacon with his other adherents were esteemed, as but *wheels agitated* by the Weight of his former and present Resentments." Although "nicely honest, affable, and without blemish in his conversations and dealings," Richard Lawrence manifested great uneasiness over the obstinate willfulness and "french Despotic Methods" of the conduct of public policy by the Governor. Colonel Richard Lee of the Council told Mathew that Lawrence "had been partially treated at Law for a Considerable Estate on behalf of a Corrupt favourite," about which he complained so loudly that Sir William "bore him a Grudge."[10]

History is full of examples of men with deep-seated grievances who embraced good causes, in part at least, to settle old scores. Richard Lawrence, William Drummond, Giles Bland, Anthony Arnold, and other leading Jamestonians looked upon Sir William Berkeley as an "Old Treacherous Villain," whose duplicity they had experienced. Together with Nathaniel Bacon they resented the Governor's penchant for conferring places, privileges, and favors on his "Grandees," men risen from lower ranks of English society.

The reform measures that these gentlemen hoped to enact were incorporated in "Nathaniel Bacon Esq'r His Manifesto Concerning the Present Troubles in Virginia." This was clearly the work of Lawrence and Drummond and probably originated in the discussions that took place on the night of June 6. It was not issued, however, until August 3 when Bacon was engaged in armed conflict with Governor Berkeley. It contained the first public expression of the contempt of the aristocrats for the parvenus.[11]

"Wee appeale to the Country itselfe what and of what nature their Oppressions have bin or by what Cabal and mistery the designes of many of those whom wee call great men have bin transacted and caryed on, but let us trace these men in Authority and Favour to whose hands the dispensation [dispensing] of the Countries wealth has bin commited; Let us observe the sudden rise of their Country Or the Reputation they have held here amongst wise and discerning men, and *lett us see wither their extractions and Education have not bin vile*, And by what pretence of learning and vertue they could soe soon [enter] into Imployments of soe great Trust and consequence, let us consider their sudden advancement and let us also consider wither any public work for o[u]r safety and defence or for the Advancem[en]t and propogation of Trade, *liberall Arts or sciences* is here Extant in any [way] adequate to o[u]r vast chardg, now let us compare these things togit[her] and see what spounges have suckt up the Publique Treasure and wither it hath not bin privately contrived away by *unworthy Favourites and juggling Parasites* whose tottering Fortunes have bin rapaired and supported at the Publique chardg, now if it be so Judg'd what greater guilt can bee then to offer to pry into these and to unriddle the misterious wiles of a powerful Cabal let all people Judge what can be of more dangerous Import then the soe long Safe proceedings of Some of o[u]r Grandees and wither People may with safety open their Eyes in soe nice a Concerne."

The "Manifesto" discloses the contours of the wide breach in the Virginia plutocracy. Bacon, Lawrence, and their fellow architects of reform, being of gentle birth and breeding, retained in the new country a lively sense of the English aristocrats' obligation to work for the public weal in return for the privileges they enjoyed—*noblesse oblige*—an obligation they charged the Ludwells and their clique of self-made tobacco planters with having neglected, repudiated, or never understood. These were times of personal politics and upper-class rule. A cleavage in the planting oligarchy brought about the contest for power between the Jamestown reformers and the

Governor's cronies that provides the key to understanding the rebellion.

The stern indictment framed by the gentlemen of Jamestown leads us to ask whether Thomas Ludwell's oft-quoted sneers about Bacon's followers did not represent a weak and malicious defensive reaction: "Now tag, rag and bobtail carry a high hand," and they are "the scum of the Country," a "Rabble of the basest sort of People." Nathaniel Bacon's closest advisers and supporters, however, were among the first gentlemen of Virginia, the best of the better sort; the rank and file of his followers were freeholders and freemen. Echoing the sentiments of Richard Lawrence, the "Rebel" himself complained that "things have been carried by men at the helm as if it were but to play a booty game, or divide a spoyle."[12]

The day after the fateful midnight conference, June 7, Nathaniel Bacon was captured and taken before the Governor and Assembly. On the advice of his friends, he submitted—quietly on bended knee—to authority and received Sir William's pardon: "God forgive you, I forgive you," repeated three times. On June 10, Bacon was restored to the Council, a move he shrewdly attributed to the Governor's desire to keep him out of the Assembly. A few days later the word went around the town: "Bacon is fled." When Thomas Mathew heard the news he "went Straight to Mr. Lawrence," who confirmed the rumor, adding that at daybreak his house had been searched by Berkeley's men. The "Rebel," writing to explain his precipitate flight and escape, reported that they were "feeling the very beds for mee," and "seized on my Servants in Town." The next thing Governor Berkeley learned was that the country people were "hastening down with Dreadfull Threatnings to double Revenge all Wrongs [that] shoud be done to Mr. Bacon or his Men." Never before had the community known such excitement.[13]

After witnessing the pardon of Nathaniel Bacon, the members of the Assembly settled down to the business of the country. Governor Berkeley told them to confine themselves to providing security from Indian depredations, warning them significantly "to beware of Two Rogues, amongst us, naming Laurance and Drummond." Mathew tells us, however, that "Some Gentlemen took this opportunity to Endeavour the Redressing several Grievances [that] the Country then Labour'd under." Henry Norwood, the treasurer and one of the Governor's closest friends, admitted to Sir Joseph Williamson in July 1676 that the people's principal complaints were of "extreme and grievous taxes," injuries suffered in the courts "through the Governor's passion, [old] age, or weakness," and the great sway of the Council over the Assembly, as well as Berkeley's "licensing some to trade with the Indians and not timely suppressing their incursions." The sequence here is significant, and there were many local grievances as well. The outcome was the passage of twenty acts that session, all of which Sir William, with some reluctance, signed into law before he dissolved the Assembly on June 25.[14]

Nathaniel Bacon, accompanied by about five hundred men had returned to Jamestown on June 23 to procure a commission to lead them on an expedition to punish the natives. Threatening the Governor and the Assembly with "fyer and sword" by a melodramatic display of force in front of the State House, during which the freemen shouted, "We will have itt, we will have itt," Bacon made his demand: "I have come for a commission against the Heathen who dayly inhumanely murder us and spill our Brethrens Blood, and noe care is taken to prevent it," ending with, "God damne my Blood, I came for a commission, and a commission I will have before I goe." The next day, June 25, a commission "calculated to the height of his own desire" was prepared and delivered to him; afterward the Assembly broke up. In the negotiations between the "General" and the Governor and Assembly, Richard Lawrence and Thomas Mathew figured prominently. Sir William had no choice but to go along with them. Moreover the authorities promised to make clear to the King that Bacon was "the only man fit in Virginia to put a stop unto the bloody resolutions of the heathen."[15]

By this turn of events, "a necessary measure of self-defense" against incursions by the natives was transformed into an open struggle for power within the planting oligarchy. Many contemporaries admitted that the master mind behind it was Richard Lawrence, ably seconded by William Drummond—both of them from Jamestown—who astutely joined highly emotional political issues to the terrifying Indian problems; and, as tribune in the front rank, they placed their fellow gentleman, Nathaniel Bacon, Jr., "the Darling of the people."[16]

The "General" and his soldiers left Jamestown on June 26 for the falls of the James to mount a campaign against the Indians. When, on July 29, he learned that the Governor had double-crossed him by again proclaiming him a rebel and attempting to raise the militia of Gloucester and Middlesex counties to oppose him, Bacon led his forces back to Middle Plantation. Failing to win support in Gloucester, Sir William fled across the bay to safety among his loyal supporters in Accomac. Meanwhile Bacon's friends and admirers among the gentry had joined him at Middle Plantation where, on July 30, Lawrence drew up two notable documents.

In the first of these, "Nathaniel Bacon, Esq'r His Manifesto

Concerning the Present Troubles in Virginia," Lawrence, as noted previously, penned a searing indictment of the greed, callousness, and incompetence of the official group surrounding the Governor. At the same time he stoutly and eloquently defended the actions of the frontiersmen and Bacon from the charge of rebellion.[17]

The second document, "The Declaration of the People," contained a bill of particulars about Sir William Berkeley that concluded with a list of nineteen persons who had been his "wicked and pernicious counsellors, aides, and assisters against the commonalty in these our cruel commotions." Given his own bitter experience, one can imagine the grim satisfaction with which Lawrence wrote the third charge against the Governor: "For having abused and rendered contemptible the Majesty of Justice, and of advancing to places of judicature scandalous and ignorant favorits." Within four days, all of these individuals were to be delivered up or surrendered; otherwise their protectors would be branded "confederates and traitors to the people" and their estates would be forfeited. "This we the Commons of Virginia do declare desiring a prime union among ourselves . . . And let not the faults of the guilty be the reproach of the innocent, or the crimes of the oppressors divide and separate us who have suffered their oppressions."[18]

Bacon published the "Manifesto" and "The Declaration of the People" at a convention attended by "most of the prime gentlemen" and freemen "in these parts" at Middle Plantation on August 3. He also submitted an oath that he insisted all of them must take. Evidently it too was the work of Lawrence; it stipulated that all acts by Bacon and the "people" were legal, but that those of the Governor and Council were ruinous to the country; and further, that those who took it would resist any forces sent over from England. Many a Virginian found it impossible to subscribe to this last provision, for it could be interpreted as condoning treason. No wonder that William Drummond was reported to have told his friend, now Colonel Lawrence: "Your sword is your commission and mine: the sword must end it."[19]

Over on the Eastern Shore, Sir William Berkeley got possession of some of Bacon's sloops by a successful ruse, and on September 7 he sailed for Jamestown to marshal forces to defeat the rebels. From his ship before the village, Berkeley issued a proclamation to the eight hundred Baconians there, promising a free and ample pardon to all who would "decline Bacon's interest" and come over to him, excepting only "one Mr. Drummond and Mr. Larance, a Collonel, and both active promoters of Bacon's designes." These two, said the chronicler, had sworn "a mere pack of untruths" about the Governor. Most of the rebel soldiers abandoned the town. Richard Lawrence went with them, forsaking "his owne Howse with all his welth and a faire Cupboard of Plate intire standing, which fell into the Governour's hands the next morning."[20]

General Bacon made a rapid march to Jamestown with a small force as soon as he got news of Berkeley's presence there. "To the general astonishment of the whole country," he "blocked up" the troops from Accomac in the town and, aware that his position was weak, he completely outsmarted the Governor by a "subtle invention": he had several parties of horsemen pick up and bring to his camp "some of the prime Gentle Women whose Husbands were in towne" and placed them dressed in white aprons in front of his own men when the defendants were on the verge of attacking his forces.[21]

The "General" may have been unsportsmanlike and lacking in gallantry (as his enemies never tired of saying afterwards), but there was a civil war on. A witty onlooker tells us "that Bacon knit more knotts by his owne head in one day, than all the hands in Towne was able to untye in a wholl weeke: While these Ladyes white Aprons became of grater force to keepe the beseiged from salleing out then his works (a pittifull trench) had strength to repell the weakest shot. . . ." No assault was attempted, and Bacon removed the ladies from danger. This was the best-acted *opera bouffe* in the history of Jamestown.[22]

Governor Berkeley next ordered his eight hundred "Accomackians" to storm Bacon's trench, but they failed ignominiously, like schoolboys "going out with heavy hearts and returning with light heels." Disgusted, Sir William fled from the capital to one of his ships in the James and sailed away to safety. As Bacon entered Jamestown on September 19, a courier brought him the news that Colonel Giles Brent, a former supporter, had changed sides and was on his way south with a thousand men and was resolved to fight him. At once the Rebel General assembled his followers, told them of Brent's strength and asked whether they were willing to face him. With "shouts and acclamations" the soldiers cheerfully disburdened themselves of all impediments "except their oaths and wenches" and marched out to meet the enemy. To prevent Governor Berkeley from returning to the capital and possibly effecting a junction with Brent, Bacon consulted his "cabinet council," which advised burning the place immediately. "Lawrence that notorious rebel . . . was the first man to set fire to Jamestown by burning his own house"; Drummond promptly followed this example at his house and then carried off the public records while Bacon fired the church. The soldiers lighted the remainder of the buildings, "saying the Rogues shou'd harbour no more there."[23]

Thus did the uprising to provide for the common defense against hostile natives and the reforming of a government unresponsive to the needs of *all* Virginians in one way or another—a worthwhile and universally popular cause—end at Jamestown on September 19, 1676. Nathaniel Bacon became ill when he was in Gloucester and died on October 26, and with the return of the Governor to the burned-out village on November 9, the rebellion collapsed. Not long afterward, the Royal Commissioners reported that of 15,000 Virginians not above five hundred were "untainted in the rebellion," and that they had found the people "sullen and obstinate."[24]

The revenge taken by Sir William Berkeley was immediate, ruthless, and final. William Drummond and Giles Bland, among others, were captured and summarily executed. "The last Account of Mr. Lawrence was from an uppermost plantation," Thomas Mathew recalled, "whence he and Four others, Desperado's with horses, pistolls, etc. March'd away in a Snow Ancle Deep, who were thought to have Cast themselves into a Branch of some River, rather than be treated like Drummond." Somehow Richard Lawrence, "the chief conspirator," escaped into oblivion. It is barely possible that in disguise he was one of a strange party brought before the selectmen of Boston in July 1678: "William Mason, Bricklayer, Charles Cleate, Dancing Master, Caesar Wheeler his servant, fidler, all at John Smith's Butcher, come from Virginia, and per [Captain] George Joy said to be in the rebellion of Nathaniel Bacon there." Or he may have been in one of the other groups against whom, a year earlier, the town of Providence was publicly warned by beat of drum "not to receive in these parts Nathaniel Bacon or his accomplices who were in arms in rebellion against his majesty in Virginia."[25]

The great cause of reform in Virginia for the benefit of all settlers of all ranks was ended for the rest of the colonial era because it could never again attract such able, committed leaders as it had during Bacon's Uprising and Lawrence's Rebellion in 1676. But it is incumbent upon us to remember what the Commissioners sent over by King Charles reported after their investigation of the revolt: "So general was the guilt of the country and the innocent so few that nothing but a general pardon could clear the breach made by the rebellion and disobedience."[26]

Part II

"James Cittie in Virginia"

VIII

The Jamestown Community:
A Conspicuous Failure

For a brief time in the fall of 1676, the freemen of Virginia had high hopes of achieving a voice in their government and that somehow their lives might consist of something more than hard work and deprivation. With the death of Nathaniel Bacon and the collapse of the rebellion, any chance of realizing these goals in their lifetimes was gone.

Virginia had always been governed "aristocratically." In the first three years after the founding, 1607-10, a "President and Council" dictated all activities, and no institutional distinction was made between the colony and its principal settlement. By all accounts, envy, dissent, and strife raged daily among the leaders; "All would be Keisars [Kaisers], none inferior to the other," while the vulgar neglected husbandry and other occupations. Under a new charter of 1609, "a more absolute government was granted, monarchically": the Virginia Company was empowered to send over a governor with complete authority to rule and defend the colony against enemy attack whether by land or by sea. Out of sheer necessity, on May 24, 1610, Sir Thomas Gates, the first lieutenant governor of Virginia who was on leave from soldiering in Holland, instituted, under the charter, a military form of government that was set forth two years later in the notorious *Lawes Divine, Morall and Martiall*. Governor Gates, Lord De la Warr, Sir Thomas Dale, and their successors enforced them strictly and rigorously to good effect until 1619, and it seems likely that a series of military governors continued, for a time at least, to administer all civil affairs until the collapse of the Virginia Company.[1]

When Sir Thomas Dale moved up the James River with 350 men in September 1611 to found the town of Henrico, he left only fifty men behind at the fort under the command of George Percy, whom he had appointed "Governor of James Town." Even though military duties occupied far more of his time than civil ones, he regarded the maintaining of a "continuall and dayly Table for Gentlemen of fashion about us" as the most onerous and costly obligation of the assignment. This independent sub-command is important as the first step in differentiating local from provincial government, but it should be noted that the choice for "Governor" was a military man.[2]

At first such an officer served at the pleasure of the Governor. Ralph Hamor mentioned in 1614 that "Master John Scarfe, Lieutenant to Captain Francis West" held that post. But in 1617 Governor Samuel Argall appointed William Powell for life to be "Captain of his guards and company, Lieutenant Governor and Commander of James Town blockhouses and people there." He was succeeded by Lieutenant, later Captain, William Pierce "now of Jamestown." In view of the disruption caused by the Indian uprising and the demise of the Virginia Company, we may conjecture that Captain Pierce went right ahead to "possess and exercise authority to command, rule, and govern . . . all the people there resident." In 1623 he was reported by no less an authority than George Sandys to be inferior to none as "an expert in the country."[3]

There were, by 1614, four principal settlements in the colony: Jamestown, Kecoughtan (Elizabeth City after 1621), Henrico, and Charles City, but none was accorded institutional status until 1618, except the tentative move in that direction with the appointment of George Percy by Governor Dale. As an integral part of the Great Charter, the Company at London instructed Governor Sir George Yeardley in November that, upon arriving in Virginia, he should erect these four "ancient" communities into "cities or boroughs." Obviously Sir Edwin Sandys and his associates intended that colonial and local government should be developed simultaneously.[4]

Ten days before Sir George landed, Governor Argall anticipated matters by proclaiming on April 7, 1619: "I hereby give leave and license for the inhabitants of Jamestown to plant as members of the Corporation and Parish of the same." He also fixed the territorial limits of the corporation to include "the whole island," as well as part of the mainland and Hog Island in the James River. This was a definite advance in separating local concerns of the village (grandly termed the chief city of the Company) from those of the colony. At the first session of the new legislative Assembly, two men, both from the military forces, sat as members representing "James city."[5]

Uncertainty about just what this "Corporation" was still bothers us today. Was it merely an area—a parish; did it have a legal institutional basis; or was it, perhaps, a local court that conducted its affairs in the manner of an English court of general sessions? The few surviving records are confusing. In a letter of 1623 to London concerning two carpenters who had been shipped over to help build an inn and repair the palisades around the village, Governor Sir Francis Wyatt observed that "there is due from the Corporacion of James City" 120 pounds of tobacco for their services, "which could not be payd this year." In November the Governor advised Captain Pierce that £10 in tobacco and corn should be "levied through the Corporation of James City" upon every planter or tradesman over sixteen years old—the total amount to be 1200 pounds of tobacco and 16 barrels of corn—as "a salary to the minister of the said corporation" and any remainder to be used for "defraying the public charge of the said corporation." It is plain that the status of the corporation at this time was such that the power to tax for local purposes remained with the Governor.[6]

As settlement spread to other parts of Virginia, "it was a great trouble for all causes to be brought to Jamestown for a trial"; consequently courts were appointed in convenient places to relieve the situation. John Pory, secretary of the Assembly, tells us nothing about the old court at Jamestown in 1622, but evidently the burden of business was too much for the justices. It was arranged that there would be monthly courts held at two other places to hear small causes.[7]

The business of the colony was still centered at Jamestown: the Governor and two or three members of his Council ordinarily resided there, and the rest of the Council met there every three months. In the years after the dissolution of the Company, the Assembly did not meet annually as originally planned but only when called up by the Governor, but the courts seemed busier than ever. An order was issued in 1631 providing that a court would be held every fortnight on Monday, with a councilor presiding, and "all of 'em take their turns." By an act of 1632, four quarterly courts were assigned to Jamestown for the handling of provincial appeals, as well as the hearing of local cases. This same year, the lower house designated the Governor and Council, or the commissioners (justices of the peace) for the monthly courts, "as the parishioners of every parish shall agree," to lay out highways.[8]

A significant step toward more systematic local government was taken in 1634 when the colony was divided into eight counties, which were to be "governed as the shires of England." James City County was one of them, and because it embraced much territory on the mainland, besides the island and its village, the county was much more extensive than the corporation of 1619. The decision, forced by the rapid growth of Virginia, meant that inevitably some of the governmental activities heretofore administered at Jamestown would be transferred to the courts of the seven other counties.[9]

The basic unit of local government was the parish, familiar to every Englishman wherever he might be. Naturally, the Company's leaders in London had parish government very much in mind when they framed the Great Charter of 1618. A properly constituted English parish vestry exercised two kinds of authority: the first was related to the ministers, the edifice, and other ecclesiastical affairs; the second was over civil matters—parish taxes, the care of the poor, morals, and other concerns of local import.

From the very first, as we know, there had been a church at Jamestown but no vestry, no churchwardens, no organized parish. The *Lawes Divine, Morall and Martiall* proclaimed by Governor Gates on May 8, 1610, stipulated that "every Minister where he is resident within the same Fort . . . Townes or Towne, shall chuse unto him foure of the most religious and better disposed [men of the community], as well to in-forme of the abuses and neglects of the people in their duties, and service to God, as also to the due reparation, and keeping of the Church handsome, and fitted with all reverent observances thereunto belonging." Moreover, every minister was required to keep "a Church Booke" of all christenings, marriages, and deaths. Here is provision made for some of the principal features of parish government: two churchwardens and, probably, two sidesmen to inquire into moral lapses and for the keeping of a parish register.[10]

Because the *Lawes* were a part of a military code, any mention of a vestry, its composition and general parish organization or popular participation was purposely omitted. The transition to representative and local institutions was understandably slow and faltering, and it was seriously interrupted, if not impeded, by the disastrous clashes with the Indians and the end of the Company's rule, 1622-24. Governor Argall must have had in mind the need, if not a demand, for reform in 1619 when he issued his proclamation establishing "the Corporation and Parish" of Jamestown almost four months before the meeting of the first legislative Assembly.

Not until March 1624 did the Assembly resolve "that there be an uniformitie in our Church as neere as may be, to the Canons in Englande, both in [substance] and Circumstance, and that all persons yeeld redie obedience unto them under pain of Censure." It also ordered each parish to provide a granary, out of which corn would be dispensed for the public's use, as determined by "the major parte of [the] Freemen," that is the whole body of the parish, the *vestry* as in England. Furthermore the churchwardens were to present all breaches of the laws against swearing and drunkenness to the county commanders or justices for judgment and punishment. In 1629, in compliance with these orders, "the church wardens for the Corporation of James City" presented Henry Soney to the court for a moral lapse.[11]

These orders clearly reveal the intention of the Virginia Company of London and the General Assembly in the colony to develop parishes similar to those at home. Already certain civil functions had been allocated to the parishes. Very soon after the abrogating of the Company's charter and the placing of Virginia under the direct control of the Crown, the members of the Council, sitting with the Governor as the general court, declared: "they have researved to themselves the right of patronage of the ministers and parishes of the fower Ancyent Buroughes" and "therefore that the parishioners of the . . . Corporation are not of themselves to elect a minister." The authority to form new parishes had passed from the Company across the ocean to the general court, which retained in its own hands the right of patronage to the church living at Jamestown.[12]

It was consequently the general court of Virginia that ordered on March 9, 1642, that notice be given to the parishioners of James City County to meet at "the chief city" for "electing of a vestry." Instead of the vestry being composed of the entire body of parishioners, it was restricted to a few of "the most sufficient and selected men," but the idea of elections was retained for the parish, which still embraced the entire county. At this time vestries received the authority to repair churches, and to choose, annually, two of their number as churchwardens, and also to have a parish clerk keep a register of marriages and deaths.[13]

A comprehensive act concerning the Church of England in Virginia was passed in March 1643. It stipulated that there should be a vestry in every parish "for the makeing of the leavies and assessements for such uses as are requisite and neces-

sary . . . [and] that the most sufficient and selected men be
chosen and joyned to the minister and churchwardens to be
that Vestrie." In 1645 the Assembly specified the use of the
procedure prevailing in England for electing the vestry: It
should be "in the power of the major part of the parishioners," and those chosen should be "such men as by pluralitie
of voices shall be thought fitt." To this extent these and other
provisions established the Church of England in Virginia. All
vestries, with one exception, could choose their own ministers; that exception was James City Parish, where the privilege
was reserved for the Governor. Otherwise local government,
lay as well as ecclesiastical, was made and regulated as much
like English parish government as was possible considering the
distance and unfamiliar conditions.[14]

After the colony was divided into counties, James City
County was sending six burgesses to the General Assembly in
1643, but two years later, when it was sending eight, the
county was limited to five burgesses, and Jamestown, as the
chief city and residence of the Governor was accorded one,
like certain centers in England. After the Civil War and the
establishment of the Commonwealth, the colonial authorities,
on March 12, 1652, surrendered Virginia to the Royal Commissioners and their fleet, which was lying before Jamestown.
No changes were made, however, in either provincial or local
governments. In March 1658, under the Protectorate, the
General Assembly resolved that all matters concerning the
vestry, such as agreements with ministers, care of the poor,
and other parish obligations, were to be "referred to their own
ordering," which meant that the people would be free to
make the decision.[15]

This latest measure was largely nullified after the accession
of Charles II to the throne. Upon Governor Berkeley's return

from England with a new commission and instructions, modifications of the Virginia parish vestry of profound social and political importance took place. Vestries, heretofore "chosen by the major part of the parish," were now, in 1661, to be limited to twelve men. What in the mother country was styled a "select vestry" was to replace the open vestry or "town meeting." Two of the twelve men were to be chosen by the minister and vestrymen to serve as churchwardens; and, portentously, when any vacancy occurred, the minister and the vestrymen selected the man to fill the place. Thus it came about that within a very short time, the twelve men—by including in their little group councilors, burgesses, sheriffs, and other officials and co-opting each other—soon became an oligarchy of planter-aristocrats. Consonant with this development, the parishioners ceased to be the vestry, and Virginia's yeomanry and the villagers of the capital were deprived of the only direct participation in politics they had previously enjoyed. In 1662 the Assembly did grant to the parishes the liberty of making by-laws by a majority vote on small matters not covered by statute.[16]

When Bacon's rebels took over Jamestown in 1676, the Assembly nullified the rules establishing the oligarchic vestries, which its members had long considered a grievance; and in the sixth of Bacon's Laws, the Assembly restored the practice of electing each vestry every three years "by a majority vote" of the freeholders and freemen of the parish. Another act fixed the boundaries of Jamestown to include "the whole island as farr as Sandy Bay" and provided that the "burgess or burgesses" of the town be elected by a majority vote of the "housekeepers, freeholders and freemen" living in the borough and paying taxes "and by none other." This act also contained a clause according to the freeholders "full power to make such good and convenient by-laws as they shall think fitt." The leading Baconians living at Jamestown, Messrs. Richard Lawrence and William Drummond, obviously promoted this provision that gave effective borough status to their small community.[17]

With the repeal of Bacon's Laws in February 1677, all Virginians lost forever their open vestries; and a small privileged class resumed rule in what the Reverend Morgan Godwyn erroneously and contemptuously dubbed "their Plebeian Juntos, the Vestries" at Jamestown and in all other parishes. As a small gesture toward reform, the Assembly voted that "six sober discreet housekeepers or freeholders" chosen by the parish were to sit with the vestry and have equal votes in the assessing of parish levies. In 1679 it was determined that two men should be selected to meet in like fashion with the justices of the county court to make by-laws.[18]

The total destruction of Jamestown by the rebels once more raised the questions of whether that spot ought to be abandoned and whether Tyndall's Point on York River in Gloucester County might not be "the most convenient place for the accommodation of the country in generall to meete att . . . for the tyme to come." The proposal met with no acceptance, but from June 1676 to April 1679, successive governors resided in the Berkeley mansion at Green Spring while the General Assembly met there and at Middle Plantation (later Williamsburg). No record of the activities of the vestry survives for this period, though some body or individual exercised its authority by rebuilding the church at Jamestown.[19]

Every agency of the government had been disrupted by the conflagration at Jamestown, and recovery, institutional as well as physical, was very gradual. In 1684 the burgesses for

James City County induced the Assembly to have the bounds of Jamestown ascertained "for the Encouragement of Building therein." Reference was made at this date to the meeting of the Assembly in the still unfinished "Court House," or the fourth State House. Some construction took place, but actually it was about fifteen years before the village was returned to anything like normal conditions. Churchmen and vestrymen of the parish petitioned the House of Burgesses for a bill enabling the parish to raise "a convenient maintenance for a minister" and to exchange the glebe lands for another more commodious tract. The low esteem to which the parish as an institution had fallen at Jamestown was made evident in 1691 when the Council refused to pass a bill permitting the justices of James City County court to make by-laws restraining hogs from running at large in their capital.[20]

Jamestown, because it never became a populous place, was never able to develop a complete society; and for the same reason, its denizens never succeeded in forming a self-governing municipality or even, for any length of time, in enjoying the privileges of an English parish. Authority was divided among the Governor, the General Assembly, the James City County court, and the parish of Jamestown. Only at Jamestown did the Governor retain the right to decide who the minister would be, and after the Restoration the parishioners lost the right to elect their own vestry. Politically, the inhabitants of the village just did not count. Things might have worked out better for all concerned had not a fire in 1698 resulted in the transfer of the capital of the Old Dominion to Middle Plantation the next year. That move settled the fate of the small community, which was inevitably forced down into obscurity and decay.

IX
The People, the Site,
and the Port

The story of Jamestown is one of continuous tragedy—
wars, disease, death, fires—but Jamestown was the first enduring English settlement in America, and as such, deserves a history. That it has not attracted a chronicler may be ascribed to
the fact that during the seventeenth century this memorable
spot—once a peninsula, then an island—was not considered by
its inhabitants to be really important. Rarely did the actual
site impress newcomers from Britain: "At my first coming
hither," John Pory complained to an English lord in 1619,
"the solitary uncouthness of this place . . . did not a little
vex me." It was not Jamestown the place, however, but rather
the strange aggregations of people who lived there and the
far greater number of souls who died there and lie in unmarked graves who give it a great and lasting significance as
an historic site.

Notwithstanding the grandiloquent name of "James Cittie,"
no permanent society ever developed in the capital of the colony. During the ninety-two years of its existence in the seventeenth century (1607-99), it was never anything but a tiny
village, or perhaps more properly, a large hamlet, and it never

matured as a stable community. In a very real sense Jamestown epitomizes the temporary quality of existence in early
Virginia; but the first Virginians were also the first Jamestonians.

The strikingly tentative nature of life and the ever-present
threat of enemy attacks that marked existence for the first
Americans are at once evident when we look back from the
present day and examine the story of the immigrants who
landed at Jamestown. In the first place, until after 1676, virtually all of the inhabitants were strangers, born across the
sea and sojourners rather than permanent residents. Most of
the newcomers tarried but a brief time and then moved on to
the tobacco plantations; the village was a mere transfer post.

During the years that Virginia was maintained and governed by the Company in London (1607-24), the size of its
base in America fluctuated markedly, rising and falling according to the vagaries of immigration and disease. From the
three vessels moored in the James River, 105 men went ashore
on May 14, 1607, to a site they deemed "a verie fit place for
the erecting of a great cittie," as the over-optimistic Captain
John Smith remembered the event. Eight months later only
thirty-eight remained alive; the entire colony was reduced to
its smallest population for the entire century. During the next
three years the inhabitants of Jamestown increased to about
350, to be its largest number during the Virginia Company's
administration. By 1616 a mere fifty men served the fort and
hamlet under Lieutenant John Sharpe. A timely warning by a
friendly Indian enabled the people to escape the ravages of the
"Massacre" of 1622; actually the numbers swelled for a time
as persons fled to the capital from the devastated country
round about. Then disease struck the settlement and the mortality was much greater than that incurred during the Indian

attacks. A census of 1624 indicates that the only permanent dwellers were local officials and their families, who totaled about 183 persons. The next year a muster listed 125 individuals in the village proper and fifty-one elsewhere on the island. Of this number, there were ten blacks.[1]

Population in the colony as a whole grew slowly until about 1650 and then rapidly until the time of Bacon's Rebellion in 1676, but "James Cittie" did not share in the increase. In 1667 its twenty houses appear to have sheltered no more than 160 permanent residents. Instead of growing in population and number of structures, a report of 1676 shows, there were only 96 people living in sixteen or eighteen houses. In September of that year, as reported earlier, Bacon's forces set fire to the town, and within a few hours not a house was left standing. When the Royal Commission arrived, there were no lodgings available for them or for the troops that came over; the latter brought the population up to more than 1000—the largest number Jamestown ever attained—and all of them had to be sheltered in tents.[2]

Three years went by before Governor Lord Culpeper gave orders to rebuild the town. Twenty years after the holocaust, the Reverend James Blair stated that although between twenty and thirty houses had been erected, only seventy-six residents held land patents. The next year, on October 31, 1698, another fire consumed the State House and prison, as well as many private dwellings. Never again did more than a handful of Virginians live on Jamestown Island.[3]

The great majority of the colonists residing at Jamestown for protracted periods during the seventeenth century came from the British Isles, nearly all of them English, a small percentage Irish. Emphasizing an occasional Dutch, German, Polish, or Italian artisan sent over by the Virginia Company in 1624, in order to enhance the prestige of this or that national group, today only serves to underline the fact that the very essence of this tiny outpost was its Englishness.[4]

Like many another colonial station established by Europeans, Jamestown was a refuge or tarrying place for lonely young men. After we make every allowance for the Virginia Company, there is no denying that it sent over a sad lot of settlers, particularly those crossing between 1607 and 1618. Surveying the history of the Old Dominion in 1708, John Oldmixon stressed, "from what small beginnings the English Colony rose to the State it is in at the present: And it cannot be imagin'd, that the first Adventurers were men of Quality and Fortune, whatever the Proprietors in England were; Men of estate would not leave their Native Country, of which the English are, of all Men, most fond, to seek an Habitation in an unknown Wilderness; and what deter'd such from going thither at first, will always deter them."[5]

In writing to Sir Edwin Sandys in 1620, John Rolfe was more candid and specific in his remarks about the low caliber of the early Jamestonians: "I speake on my owne experience for these 11 yeres, I never amongst so few, have seene so many falsehearted, envious and malicious people (yea amongst some who march in the better rank)." Alderman Johnson stated that of the thousand colonists sent out in the first twelve years, those who succumbed—four out of five—were "People for the most parte of the meanest Ranke." At times, as Governor Dale conceded in 1611, "everie man allmost laments himself of being here, and murmers at his present state, though haply he would not better it in England."[6]

The population of Jamestown was overwhelmingly male throughout the seventeenth century. Consequently there was virtually no family life, and existence, particularly for the poorer sort, was dull and crude at best. Until the latter half of the century there was no social organization other than the church. The large transient element of its population could find almost no means of recreation other than in such rough and tumble rural English pleasures as running, wrestling, and bowling. There were no "guest houses" for the hundreds of servants arriving after 1619 who very much needed shelters until they were dispersed to the plantations. A substantial sum was raised to build a "fair inn" and some bricklayers were being sought when the uprising of the natives in 1622 put an end to the project. Late in the thirties, Governor Harvey complained about the lack of an inn, but it was years before that need was met.[7]

As time passed, drinking—and from all accounts immoderate drinking—became the principal means of relaxation or recreation. Great quantities of strong waters were sent over by London merchants and dispensed in the colony at prices so exorbitant as to arouse public criticism. Numerous ordinaries had sprung up all over Virginia by 1644, when the authorities decided it was time to license them and set prices for "good and wholesome diet," beer and liquors—all payable in tobacco. Merchants were forbidden in 1646 to retail wines or other strong waters "within the Corporation of James Citte or the Island" or to exceed legally set rates and prices or "to sophisticate the same." The most popular beverages in 1658 were Spanish wines (Malaga, Canary, Sherry, "Tent" or Alicante), as well as the Portuguese vintages of Madeira and Fayal. At this time charges were set for the drinks that tavernkeepers of Jamestown might make for persons settling debts in their public rooms and also for meals at ordinaries served to both masters and their servants. Taverns, inns, and ordinaries were now fulfilling all of their historic functions by providing lodgings, diet, and public rooms for conducting business or sociable drinking.[8]

The widespread practice of innkeepers cheating on their measures for drinks, charging too much, and being too liberal in the granting of credit, as well as the idleness and debauchery observable in public houses, obliged the Assembly to take measures to curb such abuses. Twice in the sixties it moved to limit the excessive number of inns; Jamestown, because accommodations were required by many transients, as well as by members of the Assembly that met there, was excepted from the order that restricted all other counties to two licensed houses. John Oldmixon, writing about the capital, commented most favorably on its taverns and eating places available in the nineties and added that "the Humour of the Virginians to live upon Plantations" kept the seat small. Gambling was also a real concern, for all ranks of Virginians seemed to be desperately given to it: on two occasions Captain de Vries was astounded to see gambling planters putting up their servants as stakes.[9]

Not until the eighties do we find any record of organized social activity in the tobacco capital, but this is not surprising considering its small size and that many of the "better sort" preferred to live on their plantations. One of the first societies was a law club formed by the Reverend John Clayton, Henry Hartwell, William Fitzhugh, and a few others, which was to meet when the general court was in session. Fitzhugh wrote to Hartwell in 1681 to remind Mr. Clayton about providing law books and to make sure that none but "loyalists" (those who had opposed Bacon) be admitted.[10]

Five years later, Governor Lord Howard of Effingham described for William Blathwayt of London the "Cockney Feast" that "was and is annually to be held on the 23d of

April." To celebrate the first year of the coronation of King James II in Jamestown, "a very handsome appearance for this place" was arranged: a sermon preached in the church, a dinner at a tavern, bonfires, gun salutes, and a toast to the royal family. Pointing out that this was the first "society" of its kind in the village, his Lordship was glad that it had been started, particularly because "it may perhaps take many persons' thoughts a little from other designes."[11]

A man need not have been born within the sound of Bow Bells to participate in the less elaborate ceremonies at Jamestown on April 27, 1689, to honor the accession of King William and Queen Mary to the throne of England. Nathaniel Bacon the Elder, president of the Council, proclaimed their majesties at 11 a.m. amid a fanfare of trumpets, drums, and cheers followed by general drinking.[12]

Governor Francis Nicholson, a bachelor who lived in a small rented house, enjoyed sharing meals and conversation at an ordinary with members of the Assembly. To curry favor with the people, as Robert Beverley sardonically put it, he instituted "Olympic Games" on St. George's Day (April 23, 1691) at Jamestown. First and second prizes were offered "to be shot for, wrestled, played at backswords and run for, by Horse and foot . . . by the better sort of Virginians onely who are Batchelers." This was indeed the first athletic competition known to have been held in American history, and it is clear testimony to the hierarchical structure of society in the Old Dominion. No small farmers, no mariners, no servants, no black slaves, no Indians might enter, even though they were single men; only gentlemen were to contest with other gentlemen. It is comforting to posterity to learn that these well-bred bachelors sent the Governor a thank-you note applauding his generosity.[13]

The quality of life at Jamestown in the first two decades was severely limited by the absence of women and children, who, as the Earl of Southampton and other worthy gentlemen of the London Company acknowledged, were essential in forming a colony. "The Plantacion can never flourish till families be planted and the respect of wives and Children fix the people on the Soyle." In 1619 the members of the first legislative Assembly agreed: "In a newe plantation it is not knowen whether man or women be more necessary," and Thomas Nicholls, a settler, insisted in 1623 that "Women are necessary members for the Colonye" but none are here.[14]

It is one of the truisms of history that the women of any age are seldom willing pioneers. The few who came over to live at Jamestown did have a choice of husbands and soon married, often well. A man had to be very careful, however. Eleanor Sprague was haled into court in June 1624 and condemned at the next divine service "publickly before the Congregatione, [to] acknowledge her offence in contractinge her selfe to two severall men at one tyme, and, penitently Confessinge her falte, shall ask god and the Congregatione's forgiveness"—Robert Marshall appears to have been the lucky man. As a warning, the court directed that in the future the penalty for like offenses would be either a whipping or a fine "according to the qualletie of the person affendinge." As a rule the lot of the few goodwives and female servants at Jamestown, where they worked at taverns, assisted in artisans' shops, or tended kitchen gardens, was certainly more bearable than that of the women servants purchased by the planters and put to work in the fields, a situation which by 1662 had become a common practice.[15]

Most of the girls and young women emigrating to Virginia were not much better in character than the men, and more than a few of them resembled Defoe's Moll Flanders. In 1662 the Assembly found it advisable to prescribe the ducking stool for the "many brabling women" who "often slander and scandalize their neighbours."[16]

On the other hand, fortune could favor "an honest industrious woman," such as Mistress Pierce who came over in the *Blessing* and lived at Jamestown for nearly twenty years (1610-29). She married Captain William Pierce who, in 1623, became lieutenant governor and captain of James City. We learn of her advance in the world from Captain John Smith, whom she encountered on a voyage to London. In his *Generall Historie*, he stated that "she hath a Garden at Jamestown, containing three or four acres [outside the village] where in one yeare she hath gathered neere an hundred bushels of excellent figges; and that, of her owne provisions, she can keep a better house in Virginia than heere in London for 3 or 400 pounds a yeare. Yet [she] went thither with little or nothing."[17]

Mistress Pierce probably rose higher than any other woman resident of early Jamestown, but we must not overlook John Pory's contention of 1619 that "we are not the veriest beggers in the world": here in Jamestown on Sundays "goes . . . a wife of one that in England had professed the black arte, not of the scholler, but of a collier of Croydon [who] weares her rough bever hat with a faire perle hatband, and a silken suite therto correspondent."[18]

The family could never develop in Jamestown as it had in the British Isles because of the shortage of women; in that respect the Virginia capital was also unlike the contemporary towns of New England; it was a predominantly male settlement. For the whole colony, the ratio of men to women was about three to one, but on the island it must have been nearer to fifteen to one. And what made the formation of a valid society even more difficult was, as Thomas Crosfield of Oxford noted in 1620, that the plantation was "much hindred by reason of a variety of languages there among them." He was referring, of course, to the many dialects spoken by men who had just recently come from Britain and Ireland, for the King's English was as yet confined pretty much to the Thames Valley. To this confusion of tongues should be added the several Algonkian dialects spoken by numerous visiting Indians and the speech of occasional members of a Dutch ship's crew.

At any one time what might be called the permanent or regular inhabitants probably constituted no more than one-half of the population, among whom were the Governor, one or two councilors, some provincial and court clerks, three or four innkeepers, a few resident merchants, and an undetermined number of artisans in the building trades and cooperage. On the farms surrounding the village and fort were several families and some indentured servants, but these various people would not have been the ones to catch the eye of a stranger, for ordinarily they went unnoticed, even by historians.

A visitor's first impression upon landing from a ship would have been the numerous floaters he encountered on the two streets or in one of the taverns: traders who had come in from ships or from inland Virginia, planters attending the courts or sessions of the General Assembly, indentured servants at work, sea captains and their crews who had come ashore, and many men—also some women—recently imported for sale as bond servants. Also among the temporary residents were prisoners—felons and Quakers—who were lodged in the jail from time to time. Before 1622 one would frequently see Indians moving freely about, trading with the whites; and occasionally some sportive native like the twelve- or thirteen-year-old

Pocahontas, naked and turning cartwheels to amuse the on-lookers. Transients such as these may have composed 60 to 70 percent of the total, and they looked upon Jamestown "not as a place of Habitation but onlye of a short sojourninge." From 1607 to 1698, Jamestown was more of a stopping-place for rootless men than the scene of a settled society.[19]

Virginia was fated to become, and throughout most of its history to remain, a rural agricultural society. Until late in the colonial period, it could never boast of even one sizable town. Beginning as a compact, palisaded military post, James-town served as the colony's seaport, commercial center, and seat of government until 1699. Like its transient population, physical properties of the island had a look of impermanence, the result of both its location and the force of events.[20]

Following explicit instructions from the Virginia Com-pany, Captain Christopher Newport and his fellows had sailed some fifty miles up the James River and picked out a penin-sula on the north bank as their landing place. It seemed to offer security from possible Spanish attack and a suitable spot for trade with the Paspahegh and other nearby Indian tribes. A prime attraction was the existence of deep water close to the shore that ensured a suitable haven for the ships. Contrari-wise, the low-lying ground, absence of a good spring of water, and the adjacent miasmic swamps, all of which they had been warned against, were either ignored or overlooked—an oversight that had tragic consequences throughout the century.[21]

Immediately upon coming ashore, the passengers, inexperi-enced though they were and exhausted after the long voyage, were put to work: some of them to build a fort, others to "cut downe trees to make [a] place to pitch their Tents." Captain John Smith, a member of the Council, "by his owne example, good words, and faire promises," directed them in cutting reeds, binding them, and thatching crude little shelters. "I well remember," wrote Smith, "wee did hang an awning (which is an old saile) to three or four trees to shadow us from the Sunne, our walles were rales of wood, our seats unhewed trees till we cut plankes. In foule weather we shifted into an old rotten tent." This arrangement served as the Anglican church until they completed "a homely thing like a barne, set upon Cratchets, covered with rafts, sedge, and earth; so also the walls: the best of our houses [were] of like curiosity; but the most part farre much worse workmanship," and they furnished very little protection against wind or rain. All seemed to be going well until January 7, 1608, when a fire broke out and the little outpost, "being thatched with reeds," was quickly consumed, including the church and the palisade around the fort and houses.[22]

Captain Smith and Matthew Scrivener promptly saw to the rebuilding of Jamestown: the fort, three new blockhouses, a storehouse, a second church, and forty or fifty thatched houses "to keepe us warm and drye." They also dug a well in the fort to improve their water supply and erected a palisade about fifteen feet high. Inasmuch as all of this construction was carried out by 270 inexperienced men, boys, and women in only three months, one may well doubt the quality of either the carpentry or the thatching.[23]

Never during its entire existence did Jamestown appeal to many people as a desirable place in which to live and make a new home. In fact both the English authorities and the settlers seriously considered abandoning it and seeking a more salubrious and suitable location. The first instance of doubt was expressed in the Virgina Company's instructions to Sir Thomas Gates in May 1609 before he went out to the colony as lieutenant governor: the London councilors advised continuing the plantation on the peninsula "but not as your situacion or Citty, because the place is unwholesome and but in the Marish of Virginia, and to keepe it only as a fitt porte for your ships to ride before and unlade" goods brought from England. In truth Sir Thomas was urged, for better defense, to make his "principall Residence and seate" away from any navigable stream.[24]

When Gates and his party arrived at Jamestown in May 1610, their first impression of the town was that it resembled "the ruins of some auntient fortification, then that any people living might now inhabit it." The Governor next learned to his dismay that only sixty out of six hundred colonists had survived "the starving time" of the past winter; upon inspecting the fort, he "found the Pallisadoes torne downe, the Ports open, the Gates from off the hinges, [the church ruined and unfrequented], and emptie houses (which Owners' death had taken from them) rent up and burnt rather than the dwellers would step into the Woods a stones' cast from them, to fetch fire-wood; and it is true, the Indians as fast killing without, if our men stirred but beyond the bounds of their Block-house, as Famine and Pestilence did within." Thoroughly discouraged by this spectacle of desolation, human misery, and misgovernment, Gates determined to "quit the Countrye."[25]

Two weeks after his arrival, having remained on shore long enough to prevent some "ill-tempered Persons" of his company from "maliciously" burning the town, Gates sailed downstream with all of the colonists on June 7, intending to join the fishing fleet at Newfoundland and return to England. Near Mulberry Island, Gates met a "skiff" sent by Lord De la Warr, the Governor, who had just arrived in Chesapeake

Bay. Persuaded by his lordship, the old settlers and the new lot of immigrants sailed back up the James and went ashore on June 10. "If ever the hand of God appeared in action of man," the Reverend Alexander Whitaker exclaimed with puritan fervor, "it was heere most evident."[26]

Immediately Lord De la Warr instructed the colonists to "cleanse the towne," and perform all necessary kinds of work to improve it. The inhabitants did begin to repair their small houses, which had been "ready to fall on their heads," though in general his orders "had but ill success." Somewhat hyperbolically, a defender of the colony claimed that "the houses which are built are as warme and defensible against winde and weather, as if they were tiled and slated; being covered above with strong boordes, and matted round within according to the fashion of the Indians."[27]

Another year passed and a new deputy governor, Sir Thomas Dale, informed the Company that he and his Council had embarked on "the reparacion of the falling Church . . . the Storehouse, a stable for our horses, a Munition house, a Powder house, a newe well for the amending of the most unwholesome water which the old afforded." They were also prompted to go about making brick for chimneys, and to erect a sturgeon house, a blockhouse on the Back River, a smith's forge, and a shelter for cattle in the winter. All this activity signaled a new departure, for Governor Dale had, upon his arrival, found most of the residents at "their daily and usual works, bowling in the streetes."[28]

Ralph Hamor had nothing but praise for the remarkable improvements that he had witnessed during the strict regime under the *Lawes Divine, Morall and Martiall* proclaimed by Gates in 1610 and published at London in 1612. Under Dale's

rigorous military rule, 350 men were sent up the river to
found the town of Henrico in 1611, which before long was
more prosperous, larger, and better built than the original settlement. With no intention of decrying the achievements at
Henrico, there can be no question but that the departure of
these people from Jamestown must have drastically reduced
the population there and deprived the capital of its most able
workmen and artisans. Even so, we learn that in 1614 "the
Towne it selfe by the care and providence" of the officers "is
reduced into a handsome forme, and has in it two fair rowes
of howses, all of Framed Timber, two stories, and an upper
Garret, or Corne loft high, besides three large and substantiall
Storehouses, joined together in length some hundred and
twenty foot; and this town hath been lately newly, and
strongly impaled, and a fair Platforme for Ordenance in the
west Bulwarke raised; there are also without this town in the
Island some very pleasant and beautifull howses."[29]

Two years after this glowing report, John Rolfe returned
(June 1617) to find that Lieutenant Sharpe and fifty men
and boys in the fort comprised the entire population. All the
place needed, Rolfe said, was more "good and sufficient men
. . . of good birth and quality to command," as well as soldiers, artificers, laborers, and husbandmen to perform the
community's work. But they were not forthcoming, for in
May of that year, when Samuel Argall had replaced George
Yeardley as Governor, "he found but five or six houses, the
Church downe, the Palizados broken, the Bridge [or pier] in
pieces, and the well of fresh water spoiled." The storehouse
served as the church, and the market place and streets, and
any other vacant spots were planted with tobacco. It was the
old story all over again: Jamestown always lacked good
housing in the early years because unskilled craftsmen, using

poor materials, inevitably raised flimsy little structures; and their owners seldom kept them in repair.[30]

Governor Argall's instructions gave him the choice of two places for his "seat" and, because he preferred Jamestown to Bermuda Hundred forty miles up the river, once again the decaying village was saved. Samuel Argall pressed so hard for better housing that he deserves full credit for the improvements made. Rolfe declared in 1618 that we "were constrained every yeere to build and repair our old Cottages, which were always a decaying in all places of the Countrie." The Governor also saw to the erecting of the third church in 1617-19— a framed edifice measuring 50 by 20 feet—at the charge of the inhabitants.[31]

As they surveyed the twelve years of Sir Thomas Smith's government of the Virginia Company, 1607-19, the members of the Assembly admitted in 1624 that "for o[u]r houses and churches in those tymes, they were so meane and poore by reson of those calamities that they could not stand above one or two yeares, the people never going to worke but out of the bitterness of theire spirits . . . , neither could a blessinge from God be hoped for in those buildings which were founded uppon the bloud of so many Christians."[32]

A notable feature of the policy of the Virginia Company under the direction of Sir Edwin Sandys after 1619 was a plan to build a "New Towne" east of the fort, which the community had outgrown by this time. Sir George Yeardley, who was again filling the office of Governor, started implementing the project between 1619 and 1621. The colonists themselves expressed a willingness to give up the island settlement altogether; and in January 1622 the local Council asked the authorities at London to send over some skillful men to build forts, "but give us leave alsoe, to devise with them of the Most Comodius and Defencible place for the seatinge of the Chieff Cyttie of this kingdome, if they shall finde James City nott fitt or proper for that purpose."[33]

Before a decision could be made on the Council's request, the "Massacre" of March 1622 occurred, and perhaps the ability of the inhabitants to defend themselves against Indian attacks dictated the order from London of August 1622 to the Governor and Council for the continuance of Jamestown as the seat of government and recommending that "for theire better Civill government (which mutuall societie doth most conduce unto), wee think it fitt that the houses and buildings be so contrived together as may make if not hansome Towns, yet compact and orderly villages" at all new plantations. Accordingly Jamestown Island was designated one of six fortified posts.[34]

Jamestown did not prosper as the Company and Council had hoped and expected, but a new section of land was laid out for dwelling houses by William Claiborne, the surveyor general who had arrived with Governor Sir Francis Wyatt in 1621. Located on the fourth ridge east of the old stockade and named the "New Towne," it was obviously the embodiment of the plan recommended by the Company and instituted by Governor Yeardley. The "Country House" built by Governor Gates sometime in 1611, to which an ell was added by Governor Argall in 1617, became the Governor's residence. The district grew into the most substantially built and thickly populated area, and before long people were speaking of "the back street" and the "turf fort." At the same time, farms spread over the island where hogs and cattle fed on the rich marsh land along Passmore Creek and in Pitch-and-Tar Swamp.[35]

The construction of these new houses and the expansion of settlement gave rise to a critical debate. Captain Nathaniel Butler charged in 1623 that "the howses are generally the worst that I saw ever, the meanest Cottages in England beinge every way equall (if not superior) with moste of the best." To this comment the planters countered that most of the houses were "built for use and not for ornament, and are soe far from beinge soe meane as they are reported," that laboring men's houses in England are much inferior to them. As for the dwellings of "men of the better Ranke and Quallety, without blushinge [we] can make excepcion against them." Apparently the buildings of the New Town belonged exclusively to men of the better sort.[36]

There can be no doubt that housing was a prime concern of the residents of Jamestown, and the number of dwellings is an index, to some extent, of the stability of the settlement. Upon landing in 1607 the immigrants put up forty to fifty shelters, which were destroyed by fire the following January but were all replaced. The number varied from time to time, but the low point under the Virginia Company up to 1624 was probably in 1617 when Governor Argall, on his arrival, found only "five or six houses." A census of 1625 indicated that thirty-three houses were occupied by 175 persons, and in addition there were three stores or warehouses. On the neck near the village were about thirty-one farmhouses and three storehouses for tobacco and corn.[37]

John Smith described the half-urban, half-rural complexion of the place in 1629: "James Towne is yet their chiefe seat, most of the woods destroyed, little corne there planted, but all converted into pasture and gardens . . . Here most of the Cattle doe feed, their Owners being most some one way, some another, about their plantations, and returne againe when they please, or any shipping comes in to trade. Here in winter they have hay for the Cattel." This was indeed an *urbs rure*.[38]

When John Harvey assumed the governorship of Virginia in 1630, there were no inns or taverns on the island, and he was forced to play mine host in his own house to visitors, some of whom stayed a fortnight or a month while on the country's or King's service. Furthermore his residence, "a double building," became the "randezvous for all sorts of strangers," and "a generall harbour for all comers." To relieve the situation, royal officials directed Harvey to try to draw people back into the town—the first of several abortive attempts made during the century—which at that moment could be effected only by confining trade to one port.[39]

The General Assembly voted in 1635 to grant a house lot and garden plot to every person who located in the village and built a house within six months. Two years later the act was renewed, and the Governor reported to Whitehall: "There are twelve houses and stores built in the Towne, one of brick by the Secretary [Matthew Kemp], the fairest that was ever knowen in this countrye for substance and uniformitye, by whose example others have undertaken to build framed howses to beautifye the place, consonant to his majesties Instructions that wee should not suffer men to build slight cottages as heretofore." From their own purses, Harvey continued, the villagers have contributed to the erection of a brick church, and have persuaded masters of ships and planters to support the undertaking liberally.[40]

In the meantime the Assembly voted a levy for "a State House." At long last, or so it appeared, the metropolis of Vir-

ginia was taking on the attributes of a settled community—
the Governor informed their lordships that "there was not
one foot of ground half a mile together by the river's side in
Jamestown but was taken up and undertaken to be built
[upon]." George Menefie, the merchant, went over to England with part of the tobacco raised in the levy to engage
skillful workmen, but perhaps he was unable to enlist the artisans needed for the task. On April 7, 1641, Harvey con-
veyed to the colony "all that capital messuage or tenement
[with a party wall] now used for a court house." The next
year the Assembly made "a free gift" of these structures and
an orchard to the new governor, Sir William Berkeley,
whose instructions provided for the erection of a "convenient
house" for the meetings of the Council and the county and
provincial courts.[41]

To encourage the building of more substantial dwelling
houses, the new Governor's instructions also provided for
grants of 500 acres of land for each "convenient house of
brick" having a cellar and measuring 24 by 16 feet. "Because
the buildings at Jamestown are for the most part decayed and
the place found to be unhealthy and inconvenient in many
respects," their lordships further authorized the Governor,
Council, and Assembly to move the "chief town" and residence of the Governor to another location, but they were to
retain the "ancient name of Jamestown." However, the members of the General Assembly preferred the old site and but-
tressed their decision on March 2, 1643, by passing an act to
protect all persons who had built after January 1641 on lots
"long deserted" against the former proprietors, who were to
be reimbursed with tracts of equal size elsewhere.[42]

During the decade of the forties, prosperity, experience,

and the improved skills of construction workers led to the gradual evolution of a dwelling house vernacular* with some architectural merit that would become traditional both in town and country—the Assembly took note of this fact by ordering in 1647 that new county prisons be erected "according to the forme of Virginia houses," which were story-and-a-half structures, one-room deep, built of wood "high and fair," sheathed with "clove board" or clapboard. The interiors had large rooms with high ceilings, which were often "daubed and whitelined, glazed and flowered"; and if they did not have glazed windows, there were shutters "made very pretty and convenient."[43]

Governor Berkeley's order that new houses were to be made of brick was impossible to carry out in all instances; some few of them were of brick; others were cased with brick in the manner of modern veneer; but as a rule, the common Virginia house was walled with planks or clapboard, and boasted a brick chimney. Shingles took the place of tiles, which few workmen knew how to make or hang. A visitor in 1655 would have found a commingling of old thatched cottages, some pleasant framed houses of "the Virginia forme," and a few brick buildings. By all odds the most prominent of the last group was one made of three tenements, previously owned by Governor Berkeley (two of which had been presented to him upon his arrival in 1642 and the third one added later). On March 30, 1655, the Governor deeded one of these, "my house in James City," to Richard Bennett, governor of Virginia, for 27,500 pounds of tobacco.[44]

Martin Noell, a colonial expert of London, argued con-

---

* When architectural historians speak of the vernacular they mean the results achieved by people who build, as they speak, without realizing that some other mode of expression might exist and be better.

vincingly in 1662 that the Old Dominion "had subsisted under slow improvements by reason it had hitherto been a Scattered Colonie, uncollected into Bodies, Townes, and convenient Settlements of Trade, Negotiation, and Securitie, which have been the Occasion of the growing Prosperities, Strength, and Trade of New England and Barbados." He suggested that a town should be located on every great river. Encouraged by Governor Berkeley, the Assembly voted in December of that same year that "a towne be built at James Citty," it being the best spot and having the best workmen in the James valley. The village must once again have become dilapidated, for the act stipulated the erection of thirty-two brick houses, each 40 by 20 by 18 feet, that were to be placed regularly around a square. Their roofs were to be either of tile or slate.[45]

"With all chearfulnesse" the planters' Assembly contributed to the new construction at Jamestown, unanimously confessing that without such a seat "wee could not long be civill, rich, or happy, that it was the first step to our security from our Indian Enemies, and the only meanes to bring in those Commodities all wise men had so long expected from us." The Assembly called upon the Governor to require each county to build one house in Jamestown at its own charge. Had the Council been as forward-looking as the Assembly, Sir William remarked, "wee should have had a brave Towne indeed, but the poorer sort see that want and misery will sooner come upon them for want of a Towne then on the rich men," and that makes them more willing to contribute for their future good than the rich men "who are still looking back on England with hopes that the selling of what they have here will make them live plentifully there." And, he continued, "many have not been deceived in that opinion, which has been a stopp to the growth of this Country, for on it they expend no more then what is useful to them in order to their return to England."[46]

Very few of the planting gentry ever cared about the welfare and appearance of their island village, and the efforts of Restoration officials and Governor Berkeley achieved very little. Secretary Thomas Ludwell, who lived nearby in James City County, in a letter to an English official in April 1665, reported that in obedience to royal instructions "we have begun a town of brick and have allready built enough to accommodate both the publique affaires of the country and to begin a factory for the merchants" and hope to build even more. But the shrewd provincial secretary was only telling the Londoners what they wanted to hear. The Assembly had expended 251,400 pounds of tobacco levied to survey the new town, begin construction on a State House and eleven other buildings, as well as pay the bills for the bricks and lime. Sir William had collected 80,000 pounds of tobacco for the erection of eight houses besides 30,000 pounds of the leaf for the State House, which was "to be built." Secretary Ludwell himself acquired one of the three tenements in 1671, but he was a resident official who inspired few imitators.[47]

At this same time, the inhabitants were petitioning for permission "to repaire their wooden houses," and the next year they asked to be allowed to construct "out houses of Timber." To the Governor and Council, the disposition of the marshlands on Jamestown Island for common pastures forever seemed more important than improving housing and the appearance of their village. The expenditure of at least £3000 sterling by Governor Berkeley notwithstanding, Jamestown was not made habitable, for the new structures "fell down again before they were finished"—or so the grievances from Isle of Wight County stated.[48]

On September 19, 1676, Nathaniel Bacon's rebels converted "the whole town into flames, cinders, and ashes, not so much as sparing the church." The two most substantial brick dwellings were set on fire by their owners, Colonel Richard Lawrence and William Drummond, and the latter carried off the colony records after he had put the torch to the State House. As noted before, Bacon himself fired the church. The devastation was complete.[49]

Even after this disaster there was apparently no thought of forsaking the spot. The Governor, courts, and Assembly conducted the colony's business in Sir William Berkeley's nearby mansion at Green Spring. Early in 1677 the Governor told Commissioner Francis Moryson that "if there are Carpenters and Oxen or Horses to draw in the Timber, ten houses, att least, may be built in a day." In 1679 the Privy Council thought it necessary to prod the Virginia capital's officials and ordered James City rebuilt "as the most ancient and convenient place" in Virginia to be the metropolis. It was May 1680 when Governor Lord Culpeper tried to comply by urging the members of his Council to erect new brick houses in Jamestown and live in them. In June an act passed providing for the creation of a new town with a tobacco storehouse in each county and for the purchase, by the colony, of fifty acres in each for that purpose.[50]

Jamestown was partly restored as the seat and port for James City County, but creating new towns on other rivers proved nearly impossible. Representing "The Inhabitants and Freeholders of James City," William Sherwood and William Harrison petitioned the provincial government in 1682 stating that, although their town had been the seat for the courts and the "metropolis of his Majesty's Country" since 1607, no "certain limits and bounds" had ever been set. They sought

permission to erect a storehouse and authority to build anywhere on the entire island: "And itt is our desyer that all Nu-
sances and corrupcions of the Air be here after removed; and
the Citty for the future kept clean and decent, which cannot
well be done with a Law." With respect to the grants of land
stipulated in the act creating towns, they reminded the authorities that "the Land in the said Citty is of Considerable
value and not an acre there but cost above £5 sterling, besides our great charg in building."[51]

The General Assembly complied with the terms of the petition in April 1682, but whether the Governor and Council
concurred is not known. Whatever their view, the King dis-
allowed the town act, and the Virginians were left to con-
tinue their "wild and Rambling way of living." Because of
this unlooked-for veto almost no new construction took place
at Jamestown; in part, as the disillusioned Culpeper wrote
home in 1683, because some people had come to believe that
other situations in the country were more suitable for a metropolis. Inasmuch as his earlier attempt to persuade councol-
ors and "chief" inhabitants to build houses in the capital had
been fruitless, he concluded that "nothing but profit and advantage can do it, and then there will be no need of anything else."[52]

Between 1684 and 1698, the town was pretty well restored.
A brick prison "after the forme of Virginia housing" was
opened about 1685, and in 1697 Commissary James Blair reported to the Board of Trade that about twenty or thirty
houses had gone up. He also mentioned a new State House
and that the church had been rebuilt. Governor Francis
Nicholson returned in 1698 bearing instructions "to rebuild
Jamestown." On October 31, however, the third great fire

broke out and destroyed the entire hamlet. "James Cittie" never recovered from this conflagration. Ninety-two years after it was founded, Governor Nicholson persuaded the General Assembly to move the capital seven miles inland to Middle Plantation, "a healthier and more convenient place, and free from the annoyance of mosquitoes." Describing the scene of the onetime capital in 1716, the Huguenot John Fontaine said: "This town chiefly consists in a church, a Court House [rebuilt from the bricks of the former State House], and three or four brick houses . . . It was fortified with a small rampart with embrasures, but now all is gone to ruin."[53]

A primary consideration in the Virginia Company's venture into colonizing was the expectation of profits from the importation of exotic produce so much desired in England. Throughout the Company's existence from 1607 to 1624, Jamestown was the seaport and transfer point at which manufactured goods and other items entered the colony and from which local produce was shipped off to England.

Under the Company's orders in 1609, Captain Samuel Argall shaped his course from Portsmouth to the Chesapeake along a northern route, thereby avoiding pirates and the Spanish. Making his crossing in nine weeks, "he succeeded almost beyond our hopes," John Smith exclaimed, for Argall had opened a safer and quicker sea lane that would be followed by most mariners in the future. With the development of tobacco as the Virginia staple after 1612 and the increased immigration of indentured servants to work on the plantations, the port of Jamestown grew in importance as the American terminus of the transatlantic traffic. In the four years from 1619 to 1623, about forty-two sail of ships of "great burthen" arrived there, whereas the previous four years not more than twelve had come.[54]

Richard Frethorne wrote his parents on March 20, 1623, from Martin's Hundred Plantation, which was ten miles downstream from Jamestown, that "there lie all the ships that Come to the land, and there they must deliver their goodes." He and several fellow servants would sail, or more probably row, up at night in their small boats and load the next day by noon, proceed downstream in the afternoon, unload, "and then away again in the night." At Jamestown, Goodman Jackson, a gunsmith, took pity on Richard and made a cabin for him to stay in when he came up and gave him some codfish to eat with his bread and water. As the tobacco culture spread, the wearisome trips to Jamestown caused the planters to seek direct access to the great ships in the James River. At this very time the Company's officials learned that "there are divers small plantacions all seated uppon the River's sides, accessible by Boates, and before them, most of the Shipps above 200 Tun may ryde."[55]

To counter this incipient threat to the port's monopoly, the General Assembly ruled in March 1624 that ships' masters were not to break bulk or undertake private sales of their cargo without first touching at Jamestown and procuring permission from the Governor and Council to do so. After the fall of the Virginia Company in 1624, its exclusive trade to Virginia was thrown open to independent shippers, and the responsibility to supply the colony with European goods devolved upon the royal government. Under the aegis of such merchants as John Preen, who sent over four consignments of English goods between 1625 and 1628, the colonists were better supplied with the necessaries and luxuries than previously. A royal order of 1628 requiring all produce from Virginia to be carried to England tended to increase the seaport's importance.[56]

The vast waterway composed of the Chesapeake Bay and its great tributary rivers made possible the expansion of the tobacco culture throughout Tidewater Virginia, but at the same time it prevented Jamestown from ever being, for long, the chief port of the colony. Plantations appeared along the York River after 1630, and the masters of vessels began to sail directly to them for tobacco without calling at Jamestown in the adjacent river. The Assembly acted to restrict this growing practice in September 1632 by again ordering all ships arriving from England or any other parts "to sayle upp to the porte of James Citty" and forbidding their masters to break bulk before anchoring there; the penalty for violating this act was imprisonment for one month.[57]

At the same time the Assembly made it lawful for ships belonging to any inhabitants or residents coming in from the open sea to sail directly "unto any place or places to which they belonge, or at which they desire to unlade their goods." Here is one explanation of why and how so many captains of English ships acquired land in the colony: it freed them from an onerous obligation. A further act of February 1633 designated four additional places on James River where warehouses were to be built for storing tobacco awaiting shipment. Thus had the monopoly of shipping enjoyed by the merchants of Jamestown been breached.[58]

Realizing that the relaxation of shipping regulations was jeopardizing the commercial prospects of the village, the Assembly repealed the act the following year and passed a new one requiring the master of "every ship, barque, boate or vessell" from overseas to stop at the fort at Point Comfort for inspection, then proceed up the river to land his cargo at Jamestown; the storekeeper there would receive 1 percent of all tobacco paid to local merchants for the European goods newly imported. But the damage to Jamestown caused by the earlier regulations proved irreparable.[59]

The requirement of King Charles I that all commodities from Virginia be shipped directly to England was being circumvented or deliberately ignored at Jamestown by 1633; the Dutchman David Pietersen de Vries first traded there then, and other skippers from the Low Countries soon appeared. The planters liked to deal with them because they called oftener than the English did and brought more goods; and particularly because their freight charges for carrying away tobacco were lower and they allowed long-term credit—usually two years. In 1642 the Assembly decreed that it was lawful for any Dutch merchant or factor to trade in Virginia if letters from London merchants indicated that his credit was good. De Vries noted the presence of thirty ships from England in the James River in October 1644, but he also observed that those from the Netherlands "make a great trade here every year." Of twenty-four ships trading to the colony in 1647, exactly one-half of them were "Hollanders." The House of Lords was told a year later that the twenty-five Dutch ships, which were fitting out for the Virginia trade "with goods and merchandize," besides costing the English Customs £1500 annually in revenue would also mean that thirty of their ships would suffer severe losses.[60]

The determination of authorities in London to put a stop to Dutch competition caused Sir William Berkeley, in an address to the Virginia Assembly in 1651, to exclaim: "We can only feare the Londoners who would faine bring us to the same poverty, wherein the Dutch found and relieved us."

The consequences of English policy were the Navigation Act of 1651 and war with the United Provinces, 1652-54; legally the Dutch were excluded from trading in Virginia. Nevertheless some clandestine trafficking went on as late as 1662; one of the Virginians involved was William Byrd of Westover.[61]

Meanwhile efforts were still being made to confine all transatlantic shipping to the port of Jamestown. Ships with goods or servants for sale were directed in 1643 to proceed to Jamestown and ride at anchor for at least twenty-four hours upon pain of forfeiture of the goods. In 1656 the masters of all ships arriving were told to report to the Governor and to douse their sails after passing Point Comfort. Six years later the General Assembly, recognizing that these regulations imposed real hardships on the inhabitants living on other rivers, bowed to the inevitable and reduced Jamestown to the status of sole port of entry for the James River only. Thus ended fifty-five years of effort by both royal and provincial authorities to make Jamestown the exclusive port for all ships entering the Chesapeake wherever they were bound. Henceforth the privilege had to be shared with villages on other rivers, such as Hobb's Hole on the Rappahannock.[62]

Within a few years after the founding of Jamestown, authorities provided docking and other facilities to enhance its attraction to shipping and commercial interests. In 1611 Governor Dale had a bridge, or pier, built out into the James to "land our goods dry and safe." Possibly this was the same one that Governor Argall declared in 1619 to be the port's best feature, "a bridge to land goods at all time," but a year later his successor made the members of the guard-force at the fort contribute their labor to build a new wharf to replace it. The "Great Bridge" for the landing of cargoes and safety of men's lives was erected at a considerable cost, but it too was "decayed and broken down" in 1623. Sundry seamen testified in London that "generally the landing is verrye badd bothe for men and goods." The stakes or piles of the old wharves and fish weirs had become such menaces to shipping in 1663 that the townsmen were ordered to pull them up. Apparently there was an adequate number of warehouses for the storage of imported goods and storehouses for tobacco assembled in hogsheads for shipment, since they could be "built at little cost."[63]

After the Restoration of 1660, some efforts to improve the port for commercial purposes were undertaken: Captain William Owen was appointed chief pilot for the James River, and upon complaints by ships' masters, good and sufficient beacons were maintained at all dangerous places as far upriver as Jamestown. In 1665 the general court ordered that two ferryboats be kept in service between Jamestown and Surry, and by this time too, there was a forty-foot broad highway leading to the island from the interior of James City County. All of these improvements, however, were not enough to ensure a flourishing seaport. The facilities were no more suitable for handling ships and cargoes than the houses were for habitation.[64]

Efforts to increase economic activities in the village were no more successful. The one attempt to set up a weekly market at Jamestown in 1644 came to nothing, and the act was repealed in 1646. In answer to some queries from Whitehall in 1683, Governor Lord Culpeper said of markets: "there were none except a most sad one at Jamestown." Artisans were so few as to be negligible. Most craftwork had to be imported from England, for early efforts at glassmaking, the manufacturing of tiles, and the burning of bricks had lagged.

Although elsewhere in Virginia an exotic product, tobacco, was grown and shipped to England in ever mounting quantities, the sole important employment of the settlers on Jamestown Island was not growing tobacco, not shipping, but cattle-grazing and the raising of swine.[65]

X

Jamestown:
Symbol for Americans

The colony of Virginia grew and flourished after 1630, but its capital, unlike Boston, Newport, or New York, failed to develop and mature as a community or urban center. By 1675 the little outpost had begun to show signs of permanence and general improvement, if not of growing numbers; but in a matter of an hour or two Nathaniel Bacon and his men reduced it to ashes. "This unhappy town did never after arrive at the perfection it then had," Robert Beverley, one of its inhabitants and landholders, bitterly recorded in 1705, "and now it is almost deserted by the wild project of Governor Nicholson, who procured that the Assembly and General Court should be removed from thence to Williamsburg," an inland place about seven miles from it.[1]

Why, we may ask, did the settlement decline and the inhabitants desert it? And why, in view of the fact that today the only remains of the seventeenth-century capital are the ruined tower of the old church, which was fired by Nathaniel Bacon, and some foundations of buildings, do so many of our countrymen go to see the wooded island that was the scene of many tragedies and look upon it as the premier historical site in the United States?

If we compare the results with the objectives of the Virginia Company for its colony, we find that it failed in nearly every respect. The officials expected the settlers to raise their own food and send home "exotic" products to enrich the Company. The one "exotic" crop was tobacco, and the planters did ship out great quantities of it, but they grew it to the exclusion of necessary food, and the one-crop system that evolved spelled eventual hardship. An immediate result was that the settlers became dependent on the Indians for food supplies, and their assumption that they had the right to appropriate the Indians' corn, as well as their fields, led to the worsening of relations with the natives. More than once the Company stressed the need to improve these relations and to try to Christianize the Indians, but in these pages it has been only too clear that nothing was accomplished. The capital might have become a profitable trading center if the planters, who settled along the rivers, had been required to import and export through Jamestown; such trade as the Jamestonians had, however, was not enough to assure commercial pre-eminence.

The history of Jamestown from 1607 to 1699 is a somber chronicle, one unrelieved by either merriment or an attitude of warm humanity. There were some planters, merchants, and officials who acquired land and attained social position, power, and wealth; and they managed to manipulate the provincial and local governments to protect their special privileges and properties. But few of the thousands of Englishmen who came over as servants ever enjoyed real independence and security—even after they were released from their indentures. There were no social organizations to relieve the long days of work that were the lot of artisans, clerks, and laborers of the "poorer sort." Bowling and, above all, drinking were their chief recreations; when the Governor arranged for an athletic contest in 1691, only gentlemen were allowed to compete.

After the "Massacre" of 1622, and particularly after the one launched by Opechancanough in 1644, the inhabitants of Jamestown had little to fear from the Indians, but that could not be claimed for the white men who had moved westward and appropriated the lands of the natives. The English of that region became aroused at the indifference of Governor Berkeley and his advisers to their appeals for help in quelling the raids of the Indians upon what the colonists considered to be their land. They also grumbled about the inequities of the taxes they had to pay on their small farms in contrast to the levies on the plutocrats who owned large plantations.

Bacon's Rebellion began in 1676 on the Virginia frontier when the white men took the matter of defense into their own hands, but Richard Lawrence and William Drummond turned it into a campaign for political reform. The Assembly was forced to correct many political abuses, but with the collapse of the rebellion after the death of Bacon, those measures were repealed and the oligarchy regained its former special privileges and benefits.

Sir William Berkeley had been chosen Governor wholeheartedly by the Assembly after the Restoration and later confirmed by the Crown. The councilors and advisers whom he then appointed included "loyalists" who had fled to Virginia during the Great Civil War, and they were only too willing to abet the extension of royal authority over the colony and use their power to enrich themselves. In Jamestown the inhabitants lost the right to have a voice in local government even though other parishes were permitted to select their own vestries. In fact, although taking into consideration the wealthy merchants who exerted influence on the Council and Governor and who had, along with government officials, landed estates where they could enjoy better living conditions, there was little to attract ambitious men to come out from England without connections or wealth to advance their positions.

In contrast to this grim recital is a tale of the daring and fortitude and indomitable courage of the Englishmen who, in the face of obstacles and disasters that would have discouraged most men, remained in Virginia and succeeded in establishing the first enduring settlement of English America.

In spite of the aspersions cast on both the poorer sort and those "mighty dons" and "parvenus," there were able and inspired leaders who displayed concern for their fellow men and loyalty to the colony. The names of Captain John Smith, John Rolfe, Governor Dale, George Thorpe, Alexander Whitaker, Nathaniel Bacon, and Richard Lawrence come to mind, and there were others who were determined to try to find a better life in the New World for themselves and mankind. Very few of the lower ranks, as we have tried to make clear, achieved that life, but by the establishing of the legislative Assembly that, in a later century was broadened to be more representative, these Virginians made possible the development of the democratic government that attained the Company's goals and much more for future Americans.

Notes

I

1. *Travels and Works of Captain John Smith*, ed. Edward Arber and A. G. Bradley (Edinburgh, 1910), I, lxiii-lxiv; *The Jamestown Voyages under the First Charter, 1606-1609*, ed. Philip Barbour, Hakluyt Society, 2d ser., CXXVII (1969), II, 260.
2. Kenneth R. Andrews, "Christopher Newport of Limehouse, Mariner," *William and Mary Quarterly*, 3d ser., XI (1954), 28-41; and *Elizabethan Privateering: English Privateering during the Spanish War, 1585-1603* (Cambridge, Eng., 1964).
3. Andrews, "Christopher Newport," 40-41.
4. James A. Williamson, in George B. Park, *Richard Hakluyt and the English Voyages* (2d ed., New York, 1974), xiii.
5. Park, *Hakluyt*, 187; David Beers Quinn, *England and the Discovery of America 1481-1620* (New York, 1974), 451.
6. Carl Bridenbaugh, *Vexed and Troubled Englishmen, 1590-1642* (New York and Oxford, 1968), *passim*.
7. Barbour, *Voyages*, I, 77; II, 260; Smith, *Travels and Works*, I, xxxiv-xxxv; Quinn, *England and the Discovery*, 451-52.
8. Smith, *Travels and Works*, I, lxii-lxvi; Charles E. Hatch, Jr., *Jamestown Virginia: The Townsite and Its Story* (Washington: National Park Service, rev. ed., 1957), 2.

II

1. "Their great King Opechancanow (that bloody Monster upon 100 years old)," in "A Perfect Description of Virginia" [1649], *Tracts and Other Papers Relating . . . to the Colonies in North America*, ed. Peter Force (Washington, 1837), II, No. 8, p. 7.
2. "Relation" of Bartolomé Martinez, in Clifford M. Lewis and Albert J. Loomie, *The Spanish Jesuit Mission in Virginia, 1570-1572* (Chapel Hill, 1953), 155-56. (Cited hereafter as *SJM*.)
3. Martinez, "Relation," *SJM*, 156.
4. Father Louis de Oré, "Relation," *SJM*, 180.
5. "Relations" of Rogel and Carrera, *SJM*, 118, 131, 221; Michael Kenny, *The Romance of the Floridas* (Milwaukee, 1953), 148.
6. Felix Zubillaga, *La Florida. La Missión jesuitica y la colonización Española* (Rome, 1941), 349, n.9, for the royal order as cited in *SJM*, 21; L. A. Vigneras, "A Spanish Discovery of North Carolina in 1566," *North Carolina Historical Review*, XLVI (1969), 402-3.
7. Vigneras, "A Spanish Discovery," 408; Gonzales de Solís de Meras, *Pedro Menéndez de Avilés* [1567], ed. Jeanette T. Connor (Deland, 1923), 208-9.
8. Vigneras, "A Spanish Discovery," 411-13; Solís de Meras, *Menéndez*, 209; Cumarga's Report of the Expedition, Archivo Generale de Indias, Patronato 257, No. 3-2-4 (kindly supplied by Father Loomie).
9. Lewis and Loomie, *SJM*, 24-25, 118, 131, 136, 180; David B. Quinn, *North America from the Earliest Discovery to First Settlements*, New American Nation Series (New York, 1977), 279.
10. Martinez, "Relation," *SJM*, 156.
11. Quinn, *North America*, 279n.; Letter of Quirós and Segura, *SJM*, 89, 131-33.
12. *Travels and Works of Captain John Smith*, ed. Edward Arber and A. G. Bradley (Edinburgh, 1910), I, 22, 79, 81; Letter of

Quirós and Segura, Oré, "Relation," and Sacchini, "History," in *SJM*, 89, 180, 222, 227.

13. *SJM*, 92, 158, 180; William Strachey, *Historie of Travell into Virginia Britania* [1612], ed. Louis B. Wright and Virginia Freund, Hakluyt Society, 2d ser., CIII (1953), 61.

14. "Relation" of Oré, *SJM*, 108, 158-60, 181; William R. Gerard, "The Tapahanek Dialect of Virginia," *American Anthropologist*, n. s., VI (1904), 314.

15. Strachey, *Historie*, 104-5; James Mooney, "The Powhatan Confederacy," *American Anthropologist*, n. s., IX (1907), 129.

16. Frank G. Speck dates the beginnings of the Powhatan empire at *ca.* 1570, the very year of the return of Don Luis. It began with six original tribes situated within a twenty-five-mile radius of the falls of the James River (Richmond). The Pamunkey, to which Don Luis belonged, outnumbered the other five tribes taken together. "Chapters on the Ethnology of the Powhatan Tribes of Virginia," Heye Foundation, *Indian Notes and Monographs* (New York, 1919), I, No. 5, pp. 292-93, also map facing p. 227; Mooney, "Powhatan Confederacy," 135.

17. Smith, *Travels and Works*, I, 71.

18. Smith, *Travels and Works*, I, 1 (my italics), li-lii; *The Jamestown Voyages under the First Charter, 1606-1609*, ed. Philip Barbour, Hakluyt Society, 2d ser., CXXVI (1969), I, 134.

19. Smith, *Travels and Works*, I, 15-16, 181-82; II, 396.

20. Smith, *Travels and Works*, I, xxx, 141-42, 458-59.

21. Ralph Hamor, *A True Discourse of the Present State of Virginia* [1615], facsimile (Richmond, 1957), 6-10.

22. Hamor, *True Discourse*, 10-11, 53-54, 64-68; Smith, *Travels and Works*, II, 514.

23. Hamor, *True Discourse*, 2, 16, 27, 54-55, 59-60.

24. Hamor, *True Discourse*, 11-14, 56-57; *Records of the Virginia Company of London*, ed. Susan M. Kingsbury (Washington, 1906-35), IV, 117.

25. Smith, *Travels and Works*, II, 527-28; *Records Va. Co.*, IV, 118.

26. Smith, *Travels and Works*, II, 528; *Records Va. Co.*, IV, 118.
27. *Records Va. Co.*, III, 73-74; Smith, *Travels and Works*, II, 539.
28. *Records Va. Co.*, III, 228.
29. For the policy of "congregating," Indians in Mexico, see John H. Parry, *The Spanish Seaborne Empire* (London, 1966), 189, 223; *Records Va. Co.*, III, 14, 71, 107, 128; I, 147, 319.
30. William Crashaw, *A Sermon* (London, 1610), D4; Alexander Whitaker, *Good Newes from Virginia* (London, 1613), 25, 26-27.
31. *Records Va. Co.*, III, 556.
32. *Records Va. Co.*, III, 438, 549-50; IV, 128.
33. *Records Va. Co.*, III, 550; IV, 10.
34. Smith, *Travels and Works*, II, 572; *Records Va. Co.*, III, 550, (my italics); IV, 11.
35. Smith, *Travels and Works*, II, 572; *Records Va. Co.*, IV, 11. For a different interpretation of the role of Nemattanow, see J. F. Fausz and Jon Kukla, in *William and Mary Quarterly*, 3d ser., XXXIV (1977), 109, 117.
36. *Records Va. Co.*, III, 555, 612.
37. George Wyatt to his son, Sir Francis, undated [1624], *William and Mary Quarterly*, 3d ser., XXXIV (1977), 116-17.
38. *Recc⁻ ' Va. Co.*, II, 115; III, 549; Smith, *Travels and Works*, II, 586-87, 591.
39. "Good Newes from Virginia," *William and Mary Quarterly*, 3d ser., V (1948), 355-56.
40. *Records Va. Co.*, III, 613; IV, 11-12, 76.
41. Robert Beverley, *The History and Present State of Virginia*, ed. Louis B. Wright (Chapel Hill, 1947), 61-62.

III

1. Alexander Brown, *The Genesis of the United States* (Boston, 1890), I, 158, 160, 163, 166, 169.
2. William Strachey, *Historie of Travell into Virginia Britania* [1612], ed. Louis B. Wright and Virginia Freund, Hakluyt Society, 2d ser., CIII (London, 1953), 38, 122-23.

3. Ralph Hamor, *A True Discourse of the Present State of Virginia* [1615], facsimile edition (Richmond, 1957), 34; John Rolfe, *A True Relation of the State of Virginia Left by Sir Thomas Dale, Knight, in May last, 1616,* ed. J. C. Wyllie, E. L. Berkeley, Jr., and John M. Jennings (New Haven, 1951), 11-13.

4. Strachey, *Historie,* 72; Hamor, *True Discourse,* 6, 10.

5. Hamor, *True Discourse,* 10-11, 55-56, 61-62; Alexander Brown, *The First Republic of the United States* (Boston, 1898), 203; Philip L. Barbour, *Pocahontas and Her World* (Boston, 1970), 98-137.

6. William Stith, *The History of . . . Virginia* (Williamsburg, 1747), 136; Hamor, *True Discourse,* 55-56, 59-60; *Travels and Works of Captain John Smith,* ed. Edward Arber and A. G. Bradley (Edinburgh, 1910), II, 259.

7. Brown, *Genesis,* II, 639; Brown, *First Republic,* 238, 272, 278, 290, 343, 563; Smith, *Travels and Works,* II, 535; Hamor, *True Discourse,* 34; Customs Entries, London, in *American Historical Review,* XXVII (1922), 496, 520, 520n.; Philip Alexander Bruce, *Economic History of Virginia in the Seventeenth Century* (New York, 1935), II, 566.

8. Smith, *Travels and Works,* II, 534, 535.

9. Philip L. Barbour, "Pocahontas," *Notable American Women,* ed. Edward T. James (Cambridge, Mass., 1971), 79-80; Samuel Purchas, *Hakluytus Posthumus, or Purchas His Pilgrimes* (London, 1625), IV, 1774; Barbour, *Pocahontas and Her World,* 170-83, 185-89, 212; *Records of the Virginia Company of London,* ed. Susan M. Kingsbury (Washington, 1906), I, 459.

10. Brown, *Genesis,* II, 639; Brown, *First Republic,* 173-74; Hamor, *True Discourse,* 34; Jane Carson, "The Will of John Rolfe," *Virginia Magazine of History and Biography,* LVIII (1950), 58-65; Rolfe, *True Relation,* 18.

11. *Records Va. Co.,* I, 459; Hamor, *True Discourse,* 34.

12. Smith, *Travels and Works,* II, 535; *American Historical Review,* XXVII (1922), 497.

13. *Records Va. Co.,* III, 147, 504; IV, 141, 145.

14. *Records Va. Co.*, III, 221, 243, 256, 277; Lewis C. Gray, *History of Agriculture in the Southern United States to 1860* (Washington, 1933), I, 28, 29, 30, 213.

15. *Virginia Magazine*, IV (1896), 404; V (1898), 119-123, 274-277; Sir William Berkeley, *A Discourse and View of Virginia* (London, 1663), 2, 5.

16. De Vries, in New-York Historical Society, *Collections*, III, 125; William Waller Hening, *Statutes at Large of Virginia* (Philadelphia, 1823), I, 323-24.

IV

1. *Proceedings and Debates of the British Parliaments Respecting North America*, ed. Leo F. Stock (Washington, 1924), I, 343n.

2. John L. Cotter, *Archaeological Excavations at Jamestown, Virginia* (Washington, 1958), 22-23.

3. Alexander Brown, *The Genesis of the United States* (Boston, 1890), I, 167; Ralph Hamor, *A True Discourse of the Present State of Virginia* [1615], facsimile edition (Richmond, 1957), 16.

4. *Records of the Virginia Company of London*, ed. Susan M. Kingsbury (Washington, 1935), IV, 77-78; *Calendar of State Papers, Colonial, 1574-1660* (London, 1860), 231.

5. *Calendar of State Papers, Colonial, 1574-1660*, p. 231.

6. De Vries, New-York Historical Society, *Collections*, 2d ser., III, 7, 75, 77; Philip Alexander Bruce, *Economic History of Virginia in the Seventeenth Century* (New York, 1935), I, 137-38; "A Description of the Province of New Albion" [1648], in *Tracts and Other Papers Relating . . . to the Colonies in North America*, ed. Peter Force (Washington, 1836), II, No. 7, p. 5.

7. William Waller Hening, *Statutes at Large of Virginia* (Philadelphia, 1823), II, 515.

8. *Calendar of State Papers, Colonial, 1574-1660*, p. 18; *Philosophical Transactions of the Royal Society of London* (London, 1668), III, 576.

9. *Records Va. Co.*, III, 417, 455; IV, 41.

10. Sir Francis Wyatt to Va. Co., *William and Mary Quarterly*, 2d ser., VI (1926), 117; George Gardyner, *A Description of the New World, or, America, Island and Continents* (London, 1651), 99-100.

11. Edward D. Neill, *Virginia Vetusta* (Albany, 1885), 203-4; Wyndham B. Blanton, *Medicine in Virginia in the Seventeenth Century* (Richmond, 1930), 5.

12. *Records Va. Co.*, IV, 65, 174, 525; Public Record Office, London: C. O. 1: 2, p. 17, fol. 1.

13. *Journals of the House of Burgesses of Virginia, 1619-1658/9*, ed. H. R. McIlwaine (Richmond, 1915), 32; *Tyler's Quarterly Historical and Genealogical Magazine*, III (1921), 281; *Records Va. Co.*, III, 270; Gardyner, *New World*, 100.

14. Bruce, *Economic History*, I, 178; Hening, *Statutes*, I, 380.

15. *Records Va. Co.*, III, 243; Wesley F. Craven, *Red, White, and Black* (Charlottesville, 1971), 79; John C. Hotten, *Original Lists of Persons of Quality . . . and Others* (London, 1874), 224, 241, 258; Evarts B. Greene and Virginia D. Harrington, *American Population before the Federal Census of 1790* (New York, 1932), 143, 144; Neill, *Virginia Vetusta*, 116; *Virginia Magazine of History and Biography*, XI (1903-1904), 281.

16. Edward D. Neill, *Virginia Carolorum* (Albany, 1886), 187n.; *Cavaliers and Pioneers: Abstracts of Virginia Land Patents and Grants*, ed. Nell M. Nugent (Richmond, 1934), 118.

17. *Calendar of State Papers, Colonial, 1667-1680*, p. 52; Winthrop D. Jordan, *White Over Black: American Attitudes Toward the Negro, 1550-1812* (Chapel Hill, 1968), 44, 73, 75; *Virginia Magazine*, V (1899), 236-37; XXV (1917), 134, 137; *Judicial Cases Concerning American Slavery*, ed. Helen T. Catterall (Washington, 1926), I, 9, 10.

18. Carl Bridenbaugh, *Vexed and Troubled Englishmen, 1590-1642* (New York and Oxford, 1968), 420-21.

19. *Calendar of State Papers, Colonial, 1574-1660*, p. 268.

20. *Records Va. Co.*, IV, 235, 442, 473; *Minutes of the Council and General Court of Virginia, 1622-1642, 1670-1676*, ed. H. R. McIlwaine (Richmond, 1924), 53, 54.

21. Hening, *Statutes*, II, 417-18, 515; British Library: Egerton

Mss, 2395, fol. 365b; Robert Beverley, *The History and Present State of Virginia*, ed. Louis B. Wright (Chapel Hill, 1947), 271-72; *Calendar of State Papers, Colonial, 1699*, p. 261; *1668-1676*, p. 487.

22. [James Revel], "The poor unhappy transported felon's sorrowful account of his fourteen years' transportation at Virginia, in America [1657-1671]," ed. John M. Jennings, *Virginia Magazine*, LVI (1948), 180-94.

23. Hening, *Statutes*, II, 13.

24. *Records Va. Co.*, III, 231-32, 469, 666; IV, 37; *The Letters of John Chamberlain*, ed. Norman E. McClure (Philadelphia, 1939), II, 188.

25. Hening, *Statutes*, I, 138, 235, 282; Neill, *Virginia Carolorum*, 68-69, 72, 112, 116, 134, 142-43, 187n.; Nugent, *Cavaliers and Pioneers*, 2, 3, 54, 134, 161; De Vries, New-York Historical Society, *Collections*, III, 34.

26. "A Perfect Description of Virginia" [1649], in Force, *Tracts*, II, No. 8, p. 5.

27. Hening, *Statutes*, I, 127; *Records Va. Co.*, IV, 584.

28. British Library: Egerton Mss, 2395, fols., 354-55, 362-64; John Oldmixon, *The British Empire in America* (London, 1708), II, 289.

V

1. Perry Miller, "The Religious Impulse in the Founding of Virginia," *William and Mary Quarterly*, 3d ser., V (1948), 495-522; VI (1949), 24-41; Babette May Levy, "Early Puritanism in the Southern and Island Colonies," American Antiquarian Society, *Proceedings*, LXX (1960), 69-148.

2. George M. Brydon, *Virginia's Mother Church* (Richmond, 1947), 22; Samuel Eliot Morison, review in *New England Quarterly*, VII (1934), 731.

3. Harry C. Porter, "Alexander Whitaker, Apostle to Virginia," *William and Mary Quarterly*, 3d ser., XIV (1957), 339, 341n.; Alexander Brown, *The Genesis of the United States* (Boston, 1890), 499-500.

4. Ralph Hamor, *A True Discourse of the Present State of Virginia* [1615], facsimile edition (Richmond, 1957), 55, 60.

5. Hamor, *True Discourse*, 40-42, 54.

6. John Rolfe, *A True Relation of the State of Virginia Left by Sir Thomas Dale, Knight, in May last, 1616*, ed. J. C. Wyllie, E. L. Berkeley, Jr., and John M. Jennings (New Haven, 1951), 41 (italics mine).

7. William Gouge, *Of Domesticall Duties* (London, 1622), 8; Thomas Gataker, *Marriage Duties Briefly Couched Together* (London, 1620), 3; Carl Bridenbaugh, *Vexed and Troubled Englishmen, 1590-1642* (New York and Oxford, 1968), 29, 32.

8. William Waller Hening, *Statutes at Large of Virginia* (Philadelphia, 1823), I, 122.

9. Hening, *Statutes*, I, 144, 149, 261, 268-69, 374.

10. *Virginia Magazine of History and Biography*, VII (1899-1900), 374.

11. *Virginia Magazine*, XL (1932), 36-37; LIII (1945), 302-10; Cotton Mather, *Magnalia Christi Americana* (Hartford, 1853), I, 439; Thomas Hutchinson, *History of the Colony and Province of Massachusetts-Bay*, ed. Lawrence S. Mayo (Cambridge, Mass., 1936), I, 134n.; *Winthrop's Journal*, ed. James K. Hosmer, Original Narratives of Early American History (New York, 1908), II, 351-53; *Lower Norfolk Antiquary*, II, 84-85; *Historical Papers Relating to the History of the Church in Virginia*, ed. William S. Perry (Hartford, 1870), I, 1-2; *New England Historical and Genealogical Register*, XLVII (1893), 354.

12. Hening, *Statutes*, I, 532-33; Massachusetts Historical Society, *Collections*, 4th ser., IX (1871), 155.

13. *Lower Norfolk Antiquary*, II, 103; Hening, *Statutes*, II, 48, 180; *William and Mary Quarterly*, 2d ser., V (1925), 266.

14. Edward D. Neill, *Virginia Carolorum* (Albany, 1886), 285.

15. Joseph Besse, *A Collection of the Sufferings of the People Called Quakers* (London, 1753), 340, 351, 353.

16. *Calendar of State Papers, Colonial, 1661-1668* (London, 1860-), 118; William Edmondson, *Journal* (London, 1715), 69, 70.

17. *Travels and Works of Captain John Smith*, ed. Edward

Arber and A. G. Bradley (Edinburgh, 1910), II, 535; Samuel H. Yonge, *The Site of Old "James Towne,"* *1607-1698* (Richmond, 1907), 65.

18. According to the National Park Service, the ruins of the church seen today are those of two edifices: the inner foundations are of the frame building of 1617; the other brick foundations are those of the structure begun in 1639. Yonge, *Site*, 68; Charles M. Andrews, *Narratives of the Insurrections, 1675-1690*, Original Narratives of Early American History (New York, 1915), 70; *Minutes of the Council and General Court of Virginia, 1659/60-1693* (Richmond, 1914), 128, 151.

19. *Tracts and Other Papers Relating . . . to the Colonies in North America*, ed. Peter Force (Washington, 1844), III, No. 15, pp. 5-6.

20. Hening, *Statutes*, II, 517; *Calendar of State Papers, Colonial. 1681-1685* (London, 1860-), 504, 515. Recently church historians have contended that the Virginia clergymen were a far better lot than earlier Episcopal scholars said they were. I find their arguments unconvincing in the light of the original sources. But *cf.* Brydon, *Virginia's Mother Church*, vii, *et passim*.

21. Edmund and Dorothy S. Berkeley, *John Clayton: Pioneer of American Botany* (Chapel Hill, 1963), xx, xxiv, xxvii.

22. *William and Mary Quarterly*, 2d ser., I (1921), 114, 185; Berkeley, *John Clayton*, 3.

23. *William and Mary Quarterly*, 2d ser., I, 114; Berkeley, *John Clayton*, 3-127; Force, *Tracts*, III, No. 12.

24. Neill, *Virginia Carolorum*, 244; Force, *Tracts*, III, No. 15, p. 18; *Virginia Magazine*, V (1897-98), 54-55.

25. *Virginia Magazine*, XXXV (1928), 45, 49; Perry, *Historical Collections, Virginia*, I, 538.

26. Arthur L. Cross, *The Anglican Episcopate and the American Colonies* (Cambridge, Mass., 1902), 34, 90, 90n.

27. Brydon, *Virginia's Mother Church*, 280-87; *Executive Journals of the Council of Colonial Virginia*, ed. H. R. McIlwaine (Richmond, 1925), I, 716.

28. Brydon, *Virginia's Mother Church*, 315-16; *Journal of the*

House of Burgesses of Virginia, 1694-1702, ed. H. R. Mc-
Ilwaine (Richmond, 1913), 96, 98-99; *Executive Journals of
the Council,* 355-56.

VI

1. Edward Channing, *History of the United States* (New York,
1905), I, 202; Thomas Hutchinson, *History of the Colony and
Province of Massachusetts-Bay,* ed. Lawrence S. Mayo (Cambridge, Mass., 1936), I, 415.
2. *Journals of the House of Burgesses of Virginia, 1619-1658/9,*
ed. H. R. McIlwaine (Richmond, 1915), 36.
3. *Journals of the House of Burgesses,* 36; Wesley Frank Craven,
The Southern Colonies in the Seventeenth Century, 1607-1689
(Baton Rouge, 1949), 135-36.
4. "Proceedings of the Virginia Assembly, 1619," in *Narratives
of Early Virginia,* ed. Lyon G. Tyler, Original Narratives of
Early American History (New York, 1907), 249-50.
5. Charles M. Andrews, *The Colonial Period of American History* (New Haven, 1934), I, 185; Tyler, *Narratives of Virginia,* 251.
6. Tyler, *Narratives of Virginia,* 251-74.
7. *Records of the Virginia Company of London,* ed. Susan M.
Kingsbury (Washington, 1933), III, 482-84; William Waller
Hening, *Statutes at Large of Virginia* (Philadelphia, 1823),
I, 113.
8. Herbert L. Osgood, *The American Colonies in the Seventeenth Century* (New York, 1904), I, 96.
9. Andrews, *Colonial Period,* I, 190-91, 199-204; Craven, *Southern Colonies,* 164-65, 165n.; Hening, *Statutes,* I, 143, 305-6,
356.
10. Hening, *Statutes,* I, 124, 196, 244.
11. Hening, *Statutes,* I, 143.
12. Hening, *Statutes,* I, 305-6, 356.
13. Hening, *Statutes,* I, 126, 445; *Calendar of State Papers, Colonial, 1574-1660* (London, 1660), 321.
14. Captain Thomas Yong was in Jamestown in 1634 and discov-

ered that justice had been grossly abused when Harvey was Governor: "full of corruption and partiality, the richest and most powerful oppressing and swallowing up the poorer. . . ." *Narratives of Early Maryland*, ed. Clayton C. Hall, Original Narratives of Early American History (New York, 1910), 60; *Journal of the House of Burgesses*, 61.

15. Hening, *Statutes*, II, 63-64.
16. Wilcomb E. Washburn, *Virginia Under Charles I and Cromwell, 1625-1660* (Williamsburg, 1957), 29-62; *Virginia Magazine of History and Biography*, I (1893-94), 175-77.
17. Edward D. Neill, *Virginia Carolorum* (Albany, 1886), 310; *Bacon's Rebellion*, ed. John D. Neville (Jamestown, 1976), Nos. 111-12, p. 48; Sister Joan de Lourdes Leonard, "Operation Checkmate," *William and Mary Quarterly*, 3d ser., XXIV (1967), 44-74, especially pp. 63, 65, 73n., 74; Hening, *Statutes*, I, 491.
18. David S. Lovejoy, *The Glorious Revolution in America* (New York, 1972), 32-52; *Virginia Magazine*, LVI (1948), 264-66 (draft of charter).
19. Lovejoy, *Glorious Revolution*, 38-42; *Virginia Magazine*, XX (1912), 355-56; Neville, *Bacon's Rebellion*, 105, 108, 156.
20. Blathwayt Papers (Colonial Williamsburg), v, XVII, quoted by Lovejoy, *Glorious Revolution*, 51.
21. Donald H. Mugridge, "Lord Howard of Effingham Manuscripts," *Library of Congress Quarterly Journal of Acquisitions*, X (1953), 64, 65, 67, 69, 71.
22. [Durand of Dauphiné], *A Huguenot Exile in Virginia*, ed. Gilbert Chinard (New York, 1934), 110 (my italics).

VII

1. *Narratives of the Insurrections, 1675-1690*, ed. Charles M. Andrews, Original Narratives of Early American History (New York, 1915), 15-16, 52. This work contains the most important sources. A recent judicious treatment of the subject is Jane Carson's *Bacon's Rebellion, 1676-1976* (Jamestown,

1976); and highly useful additional abstracts of sources can be found in *Bacon's Rebellion* ed. John D. Neville (Jamestown, 1976).

2. British Library: Egerton Mss., 2395, fol. 550b; Wesley Frank Craven, *The Southern Colonies in the Seventeenth Century, 1607-1689* (Baton Rouge, 1949), 381-82; Andrews, *Narratives,* 20; Neville, *Bacon's Rebellion,* 292.

3. According to one of the most acceptable of the post-rebellion accounts, "Larance was late one of the Assembley, and a Burgis for [James] Towne, in which he was a liver. He was a Parson [person] not meanely acquainted with such learning (besides his naturall parts) that inables a Man for the Management of more then ordinary imployments, Which he subjected to an eclips, as well in the transactings of the present affaires, as in the dark imbraces of a Blackamoore, his slave: And that in so fond a Maner, as though Venus was chiefly to be worshiped in the Image of a Negro, or that Buty consisted all together in the Antiphety of Complections: to the noe meane Scandle and affront of all the Vottrises in or about towne." Whether or not there was any truth in this piece of gossip circulated after the collapse of the insurrection, when all hands hastened to defame the losers, cannot now be determined. Such conduct, however, was not unique among the planting gentry of Virginia, and certainly the governor's faction could have proved no monopoly of virtue. Neville, *Bacon's Rebellion,* 293, 294, 318; Andrews, *Narratives,* 19-20, 27, 40-41, 96, 109-10; *Cavaliers and Pioneers: Abstracts of Virginia Land Grants and Patents, 1623-1800,* ed. Nell M. Nugent (Richmond, 1934), 168.

4. Andrews, *Narratives,* 20-21, 109-10.

5. The first Assembly since 1661 had met briefly in March to deal with the Indian troubles. It raised taxes substantially. Neville, *Bacon's Rebellion,* 294; Andrews, *Narratives,* 52, 110-13.

6. Andrews, *Narratives,* 22, 24n., 38, 40, 96, 114, 145, 151-52n.; Neville, *Bacon's Rebellion,* 274, 296.

7. On or about December 12, 1673, some people in Surry gath-

ered at Lawne's Creek Church to protest against recently announced excessive taxes; later in an "old field" near Smith's Fort they met again. Only after they admitted their fault did the Governor remit the fines levied upon them. Sir William stated that he had "appeased" two mutinies in 1674, "raised by some secret villains who whispered among the people that there was nothing intended by the fifty pounds [tax levy in tobacco] but the enriching of some few people." Thus it appears that there must have been three rather than two "uprisings" before the outbreak of Bacon's Rebellion. Surry County Records, Jan. 3, 1673/4, cited by Richard L. Morton, *Colonial Virginia* (Chapel Hill, 1960), I, 225-26; and for New Kent, Thomas J. Wertenbaker, *Torchbearer of the Revolution* (Princeton, 1941), 33; Bland in *Virginia Magazine of History and Biography*, XX (1912), 354; Neville, *Bacon's Rebellion*, 45.

8. Andrews, *Narratives*, 40, 41 (my italics), 96.
9. *Virginia Magazine*, XX, 354 (my italics); Herbert L. Osgood, *The American Colonies in the Seventeenth Century* (New York, 1907), III, 217-18.
10. Andrews, *Narratives*, 27, 40.
11. Andrews, *Narratives*, 23, 24, 96; the "Manifesto" is printed in *Virginia Magazine*, I (1893-94), 56-57 (my italics).
12. *Virginia Magazine*, I, 79, 183; Coventry Papers (Longleat), Library of Congress microfilm, LXXVIII, 112, as quoted by Edmund S. Morgan, *American Slavery, American Freedom* (New York, 1975), 267; Andrews, *Narratives*, 61.
13. Andrews, *Narratives*, 23, 27; *William and Mary Quarterly* (1), IX, 9.
14. Andrews, *Narratives*, 23-24; Neville, *Bacon's Rebellion*, 293, 299, 300, 301, 319, 320; *Calendar of State Papers, Colonial, 1661-1668* (London, 1860), 484; William Waller Hening, *Statutes at Large of Virginia* (Philadelphia, 1823), II, 341-65 (Bacon's Laws), 380-81; Craven, *Southern Colonies*, 360-93.
15. *Virginia Magazine*, I, 172, 183-84; Andrews, *Narratives*, 23-30, 116-17.
16. The author of "Bacon's and Ingram's Rebellion," a dependable

contemporary, described Bacon and Drummond as "the cheife Incendyaries, and promoters to and for Bacon's Designes; . . . by whose Councells all transactions were, for the grater part, managed all along that Side." Governor Berkeley showed that he so considered them by exempting the two from his proclamation of pardon in June 1676 and thereafter. Andrews, *Narratives*, 23-24, 30, 55, 95-96.

17. *Virginia Magazine*, I, 56-57.
18. *Virginia Magazine*, I, 55-61.
19. Drummond to Lawrence: George Bancroft had access to papers at Richmond later burned in the Civil War, *History of the United States* (Boston, 1856), II, 226. For Bacon's oath, see Andrews, *Narratives*, 122; and Neville, *Bacon's Rebellion*, 299, 302.
20. Andrews, *Narratives*, 66-67.
21. Andrews, *Narratives*, 68-70, 135.
22. Andrews, *Narratives*, 69.
23. Andrews, *Narratives*, 35, 70-71, 135-36, 140.
24. *Calendar of State Papers, Colonial, 1677-1680*, pp. 19, 43, 361; *Boston Records from 1660 to 1701*, 7th Report of the Record Commissioners (Boston, 1881), 121-22; *Early Records of the Town of Providence* (Providence, 1892), VIII, 15-16.

VIII

1. John Rolfe, *A True Relation of the State of Virginia Left by Sir Thomas Dale, Knight, in May last, 1616*, ed. J. C. Wyllie, E. L. Berkeley, Jr., and John M. Jennings (New Haven, 1951), 31-35; William Waller Hening, *Statutes at Large of Virginia* (Philadelphia, 1823), I, 80-98; *The Three Charters of the Virginia Company of London*, ed. Samuel M. Bemis (Williamsburg, 1957), 27-54.
2. Edward D. Neill, *Virginia Vetusta* (Albany, 1885), 85; Charles E. Hatch, Jr., *Jamestown, Virginia: The Townsite and Its Story* (Washington: National Park Service, rev. ed., 1957), 15.

3. Ralph Hamor, *A True Discourse of the Present State of Virginia* [1615], facsimile edition (Richmond, 1957), 33; *Records of the Virginia Company of London*, ed. Susan M. Kingsbury (Washington, 1933-35), III, 87, 169; IV, 111, 209.

4. Hening, *Statutes*, I, 115; *Records Va. Co.*, III, 98-109.

5. Alexander Brown, *The First Republic of the United States* (Boston, 1898), 287; *Records Va. Co.*, III, 153.

6. *Records Va. Co.*, IV, 104, 401-2.

7. *Travels and Works of Captain John Smith*, ed. Edward Arber and A. G. Bradley (Edinburgh, 1910), II, 571; Hening, *Statutes*, I, 125.

8. Smith, *Travels and Works*, II, 885; Hening, *Statutes*, I, 168-69, 174, 199, 552.

9. Hening, *Statutes*, I, 224.

10. "For the Colony in Virginia Britannia. Lawes Divine, Morall and Martiall, &c." [1612], in *Tracts and Other Papers Relating . . . to the Colonies in North America*, ed. Peter Force (Washington, 1844), III, No. 2, p. 11.

11. *Records Va. Co.*, IV, 580-83; Hening, *Statutes*, I, 122-27; *Minutes of the Council and General Court of Virginia, 1622-1632, 1670-1676*, ed. H. R. McIlwaine (Richmond, 1924), 200.

12. *Minutes of the Council*, 22.

13. *Minutes of the Council*, 449; Hening, *Statutes*, I, 227.

14. Hening, *Statutes*, I, 240, 241-42, 290-91, 309-10.

15. Hening, *Statutes*, I, 299-300, 322, 362-63, 433, 478-79; II, 20, 106.

16. Hening, *Statutes*, II, 25, 44-45, 171-72.

17. Hening, *Statutes*, II, 362; *Narratives of the Insurrections, 1675-1690*, ed. Charles M. Andrews, Original Narratives of Early American History (New York, 1915), 35.

18. Hening, *Statutes*, II, 380-81, 396, 441; Morgan Godwyn, *The Negro's and the Indian's Advocate* (London, 1680), 168.

19. Hening, *Statutes*, II, 362, 405, 432; *Historical Collections Relating to the American Colonial Church, I, Virginia*, ed. Wiliam Stevens Perry (Hartford, 1870), I, 2.

20. *Journals of the House of Burgesses, 1659/60-1693*, ed. H. R. McIlwaine (Richmond, 1914), 251, 344; *1694-1702* (Richmond, 1913), 22, 33.

IX

1. *Travels and Works of Captain John Smith*, ed. Edward Arber and A. G. Bradley (Edinburgh, 1910), I, 6; Evarts B. Greene and Virginia D. Harrington, *American Population Before the Federal Census of 1790* (New York, 1932), 134; Edward D. Neill, *Virginia Vetusta* (Albany, 1885), 58-59; J. C. Hotten, *Original Lists of Persons of Quality . . . and Others* (London, 1874), 168-87; Irene S. Hecht, "The Virginia Muster of 1624/5," *William and Mary Quarterly*, 3d ser., XXX (1973), 73, 75, 78.

2. *Narratives of the Insurrections, 1675-1690*, ed. Charles M. Andrews, Original Narratives of Early American History (New York, 1915), 70.

3. *Journals of the House of Burgesses of Virginia, 1694-1702*, ed. H. R. McIlwaine (Richmond, 1913), xxx; Lyon G. Tyler, *The Cradle of the Republic* (2d ed., Richmond, 1906), 81.

4. See for example the recent attempt to make Dr. Lawrence Bohun of Norman descent (Bohun-Boone) and English birth a Hungarian. UPI dispatch from Richmond, Va., *Providence Sunday Journal*, December 26, 1976, B-18; and *cf.* Alexander Brown, *The Genesis of the United States* (Boston, 1890), II, 830.

5. John Oldmixon, *The British Empire in America* (London, 1708), II, 288-89.

6. *Records of the Virginia Company of London*, ed. Susan W. Kingsbury (Washington, 1933-35), III, 22, 247; IV, 4; Brown, *Genesis*, I, 506.

7. *Records Va. Co.*, III, 302; *Journals of the House of Burgesses of Virginia, 1619-1658/9*, ed. H. R. McIlwaine (Richmond, 1915), 17, 23.

8. *Records Va. Co.*, IV, 453; William Waller Hening, *Statutes at Large of Virginia* (Philadelphia, 1823), I, 287, 319, 411, 446, 489-90.

9. Hening, *Statutes*, II, 112-13, 268-69, 393-94; III, 44, 46; Oldmixon, *British Empire*, I, 272-73; De Vries, New-York Historical Society, *Collections*, 2d ser., III, 36, 125; [Durand of

Dauphiné], *A Huguenot Exile in Virginia*, ed. Gilbert Chinard (New York, 1934), 148.

10. Philip Alexander Bruce, *Institutional History of Virginia in the Seventeenth Century* (New York, 1910), I, 568-69.

11. *William and Mary Quarterly*, 2d ser., IV (1947), 327; XVI (1960), 378-84; Blathwayt Papers (Colonial Williamsburg, Va.), XIV, No. 3.

12. *Executive Journals of the Council of Colonial Virginia, 1680-1739*, ed. H. R. McIlwaine (Richmond, 1925), I, 106.

13. *William and Mary Quarterly*, (1) XI (1902), 87; Robert Beverley, *History and Present State of Virginia*, ed. Louis B. Wright (Chapel Hill, 1947), 98; Beverley Fleet, *Virginia Colonial Abstracts, 1652-1820* (Richmond, 1938-40), XXX, 3.

14. *Records Va. Co.*, I, 256, 566; III, 160, 493-94, 505; IV, 231; Oldmixon, *British Empire*, II, 289; *Calendar of State Papers, Colonial, 1574-1660* (London, 1860), 19.

15. *Minutes of the Council and General Court of Virginia, 1622-1632, 1670-1676*, ed. H. R. McIlwaine (Richmond, 1924), 15, 17; "Leah and Rachel" [1656], in *Tracts and Other Papers Relating . . . to the Colonies in North America*, ed. Peter Force (Washington, 1836-47), III, No. 14, p. 12; Hening, *Statutes*, II, 170.

16. Hening, *Statutes*, I, 166-67.

17. Smith, *Travels and Works*, II, 887; Tyler, *Cradle*, 45, 46; *Records Va. Co.*, IV, 108, 111, 209, 250, 584.

18. *Records Va. Co.*, III, 220.

19. *Diary of Thomas Crosfield*, ed. Frederick J. Boas (London, 1935), 19; Brown, *Genesis*, I, 499; II, 702n.; *Records Va. Co.*, I, 566.

20. Philip Barbour, "Pocahontas," *Notable American Women*, ed. Edward T. James (Cambridge, Mass., 1971), III, 78-79; William Strachey, *Historie of Travell into Virginia Britania* [1612], ed. Louis B. Wright and Virginia Freund, Hakluyt Society, 2d ser., CIII (1953), 62, 72.

21. Lewis C. Gray, *History of Agriculture in the Southern United States to 1860* (Washington, 1933), I, 377; John L. Cotter, *Archaeological Excavations at Jamestown, Virginia*

(Washington: National Park Service, 1958); Ivor Noël Hume, *Here Lies Virginia* (New York, 1963).

22. Smith, *Travels and Works*, I, xxxiii-iv; II, 387, 392, 612, 957; *The Jamestown Voyages under the First Charter, 1601-1609*, ed. Philip Barbour, Hakluyt Society, 2d ser., CXXVII (1969), 394.

23. Smith, *Travels and Works*, I, 154; II, 612; Brown, *Genesis*, I, 165-66, 175.

24. *Records Va. Co.*, III, 13, 16.

25. Brown, *Genesis*, I, 405; Samuel Purchas, *Hakluytus Posthumus, or Purchas His Pilgrimes* (London, 1625), IV, 1748; George Percy, "A Trewe Relacyon of the Proceedinges and Occurrences of Momente . . ." *Tyler's Quarterly Historical Magazine*, III (1921), 269; Alexander Whitaker, *Good Newes from Virginia* (London, 1613), Epis, Ded., B-1 v.

26. Purchas, *Pilgrimes*, IV, 1750-51; Brown, *Genesis*, I, 414; Whitaker, *Good Newes*, Epis. Ded., B-1 v.

27. Brown, *Genesis*, I, 415; "A True Declaration of the Estate and Colony of Virginia" [1610], Force, *Tracts*, III, No. 1, p. 20; Purchas, *Pilgrimes*, IV, 1753.

28. Neill, *Virginia Vetusta*, 81; Ralph Hamor, *A True Discourse of the Present State of Virginia* [1615], facsimile edition (Richmond, 1957), 21, 26, 29-30.

29. Hamor, *True Discourse*, 29-30, 32-34; Purchas, *Pilgrimes*, IV, 1752-53.

30. John Rolfe, *A True Relation of the State of Virginia Left by Sir Thomas Dale, Knight, in May last, 1616*, ed. J. C. Wyllie, E. L. Berkeley, and John M. Jennings (New Haven, 1951), II, 535-36.

31. *Records Va. Co.*, III, 71, 73; Smith, *Travels and Works*, II, 535-36.

32. *Journals of the House of Burgesses of Virginia, 1619-1658/9*, 21.

33. *Records Va. Co.*, III, 583; Tyler, *Cradle*, 41; Samuel H. Yonge, *The Site of Old "James Towne," 1607-1698* (Richmond, 1907), 65.

34. Yonge, *Site*, 50; *Records Va. Co.*, III, 583, 669.

35. Tyler, *Cradle*, 43, 44, 50-51; Yonge, *Site*, 52.
36. *Records Va. Co.*, II, 383; *Journals of the House of Burgesses, 1619-1658/9*, p. 24; Edward D. Neill, *History of the Virginia Company of London* (Albany, 1869), 398-99.
37. Smith, *Travels and Works*, II, 535, 887.
38. *Virginia Magazine of History and Biography*, VII (1899-1900), 364; *Records Va. Co.*, IV, 78; Henry C. Forman, *Jamestown and St. Mary's* (Baltimore, 1938), map, pp. 63-64.
39. Letter of Captain Yonge, Massachusetts Historical Society, *Collections*, 4th ser., IX (1871), 112-13; *Virginia Magazine*, III (1895-96), 29.
40. Philip Alexander Bruce, *Economic History of Virginia in the Seventeenth Century* (New York, 1935), II, 534; *Calendar of State Papers, Colonial, 1574-1660*, pp. 268, 288; *Virginia Magazine*, III, 29-30; Hening, *Statutes*, I, 267.
41. Tyler, *Cradle*, 57; *Virginia Magazine*, II (1894-95), 284-85; Hening, *Statutes*, I, 267.
42. Hening, *Statutes*, I, 252; *Virginia Magazine*, II, 284, 287.
43. Hening, *Statutes*, II, 76-77; *Narratives of Early Maryland*, ed. Clayton C. Hall, Original Narratives of Early American History (New York, 1910), 297-98: William Bullock, *Virginia Impartially Examined* (London, 1649), 3.
44. Hening, *Statutes*, I, 407; Force, *Tracts*, II, No. 7, 4; No. 8, 7; No. 14, 18.
45. British Library: Egerton Mss., 2395, fol. 666; *Virginia Magazine*, III (1895-96), 16; Hening, *Statutes*, II, 172-73.
46. British Library: Egerton Mss., 2395, fols. 262-64.
47. Public Record Office, London: C. O. 1/19, fol. 47; Edmund S. Morgan, *American Slavery, American Freedom* (New York, 1975), 190, quoting Clarendon Mss., fol. 275, in Bodleian Library, Oxford; *Calendar of State Papers, Colonial, 1661-1668* (London, 1860-   ), 533.
48. *Journal of the House of Burgesses of Virginia, 1660-1693*, ed. H. R. McIlwaine (Richmond, 1914), 56, 58, 102; Tyler, *Cradle*, 66, from Randolph Mss.; *Aspinwall Papers*, Massachusetts Historical Society, *Collections*, 4th ser., IX, 164.
49. Andrews, *Insurrections*, 70, 136, 136n.; Public Record Office: C. O. 5: 1371, p. 93.

50. Public Record Office: C. O. 5: 1371, p. 63; Royal Instruction for Lord Culpeper, March 1679, Macdonald Papers (Virginia State Library), V, 165; Hening, *Statutes*, II, 472, 475, 476.

51. Petition of "The Inhabitants & Freeholders of James Citty," 1682, Ambler Papers (Library of Congress), photostat in Virginia Historical Society.

52. *Virginia Magazine*, XXIV (1917), 144; *Calendar of State Papers, Colonial, 1681-1685*, pp. 497, 569; Yonge, *Site*, 41; Tyler, *Cradle*, 75.

53. Hening, *Statutes*, III, 15, 197, 312, 471; Tyler, *Cradle*, 73, 81; *Journals of the House of Burgesses of Virginia, 1694-1702*, xxx; Hugh Jones, *The Present State of Virginia* [1724], ed. Richard L. Morton (Chapel Hill, 1956), 25; Ann Maury, *Memoirs of a Huguenot Family* (rev. edn., New York, 1872), 271.

54. Brown, *Genesis*, I, 343; *Records Va. Co.*, II, 350, 351; III, 685.

55. *Records Va. Co.*, IV, 59, 259.

56. Hening, *Statutes*, I, 126; Arthur P. Middleton, *Tobacco Coast: A Maritime History of Chesapeake Bay in the Colonial Era* (Newport News, 1953), 249.

57. Fairfax Harrison, *Landmarks of Old Prince William* (Richmond, 1924), I, 63; Massachusetts Historical Society, *Collections*, 4th ser., IX, 111; *Calendar of State Papers, Colonial, 1574-1660*, p. 288; Hening, *Statutes*, I, 163, 166-67, 191-92, 204, 206, 257; Force, *Tracts*, III, No. 12, p. 11.

58. Hening, *Statutes*, I, 191-92.

59. Hening, *Statutes*, I, 166-67, 214-15, 221.

60. *Virginia Magazine*, VII (1899-1900), 267; I (1893-94), 77; De Vries, New-York Historical Society, *Collections*, III (1857), 37, 77-78, 124, 125; *Narratives of New Netherland*, ed. J. Franklin Jameson, Original Narratives of Early American History (New York, 1909), 195; Hening, *Statutes*, I, 258, 296; Force, *Tracts*, II, No. 8, pp. 5, 7, 14; *Proceedings and Debates of the British Parliaments respecting North America*, ed. Leo F. Stock (Washington, 1924), I, 205.

61. *Virginia Magazine*, I (1893-1894), 77; Fleet, *Virginia Colonial Abstracts, 1652-1820*, XI, 13; Holland Society of New York, *Year Book*, XIII, 132, 133, 135, 146-47, 149.

62. Hening, *Statutes*, I, 245-46, 392, 412, 414, 476; II, 35; III, 53, 58-60.

63. *Journals of the House of Burgesses, 1619-1658/9*, p. 32; Tyler, *Cradle*, 38; Alexander Brown, *The First Republic of the United States* (Boston, 1898), 254; *Records Va. Co.*, II, 302; IV, 93, 174, 178; Hening, *Statutes*, II, 103, 174, 205.

64. *Journals of the House of Burgesses, 1660-1693*, p. 11; *Minutes of the Council and General Court*, 508, 509; Hening, *Statutes*, II, 12, 103.

65. Hening, *Statutes*, I, 362, 397, 412, 476; *Records Va. Co.*, III, 586; IV, 23; *Journals of the House of Burgesses, 1619-1658/59*, p. 125; *Virginia Magazine*, V (1897-98), 41-42; LVIII (1950), 16-39; Hume, *Here Lies Virginia*, 204, 215; British Library: Additional Mss., 28218, fol. 14.

X

1. Robert Beverley, *The History and Present State of Virginia*, ed. Louis B. Wright (Chapel Hill, 1947), 86.

Appendix I

Jamestown Chronology, 1544-1699*

The Indian Period

Ca. 1544   Probable year of birth of Opechancanough.

1561   Menéndez took an Indian youth to Spain, where the King had him placed under the tutelage of the Dominicans.

1563   Now *Don Luis de Velasco*, the youth sailed with Menéndez for Mexico, and continued under Dominican auspices for three more years.

1566   Early in the year Don Luis traveled to Florida with two Dominican friars and joined a colonizing expedition to Bahía de Santa María in August. Missing the entrance to Chesapeake Bay, the ship was headed for Spain and arrived at Cadiz, October 23.

1567   In the autumn, Don Luis proposed to the Jesuits that he would lead a party of missionaries to Ajacán (his tribal country) and as a Christian assist

---

* For the colony in general, see William W. Abbot, *A Virginia Chronology, 1585-1783* (Williamsburg, 1957); for the Governors, consult Appendix IV of this work.

177

with the conversion of the natives. In November he arrived at Havana with Menéndez.

1570   Don Luis and the Jesuits reached Bahía de Santa María on September 20 and founded a mission not far from the site of the later Jamestown. Following a public reprimand, Don Luis forsook the mission and took up residence with his brothers.

1571   In February, Don Luis and other Indians slew the Jesuits—his *first* massacre. Rejecting Christianity and "going native," Don Luis took the name *Opechancanough* and assisted his brother in forming "the Powhatan Empire."

1585   Beginning of Sir Walter Raleigh's colony at Roanoke, June 24.

—1590   Before this date, some settlers of "the lost colony" of Roanoke moved northward to live along Lynnhaven Bay adjacent to the Chesapeake Indians.

1603   An English ship visited Chesapeake Bay and took away several natives when it returned to London.

1607   Probably early in April, Powhatan's braves slaughtered the entire Chesapeake tribe and some of the whites of "the lost colony"—Opechancanough's *second* massacre.

The English Period

1607   Captain Christopher Newport's three ships entered the Virginia capes; Powhatan's Indians attacked an English shore party. On May 4 alert native lookouts saw a shallop sail up the James River past Paspahegh (Jamestown Island). On May 14 the

English landed 105 men to begin building James-town. Toward the end of the month the leaders of the whites first met Opechancanough and his tribesmen. The chieftain captured Captain Smith early in December and prevented his followers from killing him. *Ca.* December 29, Smith first met Powhatan, whose daughter Pocahontas saved the soldier's life.

1608 Jamestown was almost entirely destroyed by fire on January 7; the same month Captain John Smith publicly humiliated Opechancanough and his tribesmen.

1609 A second charter issued May 23 enabled the Vir-ginia Company to assume direct administration of the colony. On October 1 two English women ar-rived at Jamestown and the first white marriage in Virginia was celebrated.

1609-10 The winter of "the starving time."

1610 Sir Thomas Gates arrived, May 23. He instituted military rule under a code: *Lawes Divine, Morall and Martiall* (London, 1612). Gates and his colo-nists abandoned Jamestown on June 17 and sailed away. Three days later Lord De la Warr and Gates returned with the rest of the English settlers.

1611 The second church and a wharf were erected.

1611-12 John Rolfe experimented successfully with West Indian tobacco.

1613 Captain Samuel Argall captured Pocahontas and brought her to Jamestown as a hostage on April 13.

1614 The English forced Powhatan to make peace in March. On April 5 John Rolfe and Pocahontas were married in the Jamestown church. Rolfe shipped the first cargo of tobacco to England June 28.

1614-22 By secret diplomacy Opechancanough built up the military strength of the Powhatan empire and lulled the English into a false sense of security.

1616 John and Rebecca Rolfe crossed to London with Governor Sir Thomas Dale and a party of Indians, including Tomocomo, in May. Late that summer Governor George Yeardley and his soldiers slaughtered between twenty-four and forty Chickahominy Indians needlessly.

1617-19 Governor Argall's new church under construction.

1617 Yeardley's promotion of tobacco as a staple succeeded. Rolfe returned a widower after the death of Pocahontas on March 21.

1618 Powhatan died in April; succeeded by his brother, Itopatin. Argall proclaimed Jamestown "a corporation and parish" on April 7, and on April 20 Governor Yeardley freed the servants of the Virginia Company.

1619 During this year Opechancanough replaced Itopatin as *the Powhatan*.

1619-21 Yeardley began the construction of "the New Towne"; Opechancanough secretly prepared to exterminate all Englishmen.

1619 July 30 to August 4, the first legislative Assembly in America met in the church at Jamestown. In August a Dutch ship brought the first blacks to Virginia.

1621 In August the Virginia Company began to send women to Virginia.

1622 Early in March, Jack of the Feathers was killed. On March 22 Opechancanough's Indians fell upon the settlements and killed nearly five hundred colonists—the *third* massacre. March to December, the whites inflicted a terrible revenge upon the natives; severe mortality from disease exceeded deaths in the "Massacre."

1624 The Assembly ordered religious uniformity according to the Church of England in May, and this same month the Virginia Company lost its charter. Virginia became a royal province, and trade at Jamestown was opened to all English ships.

1632 September: quarterly courts were established at Jamestown; all ships were ordered to enter and clear at this tiny port. Indian war finally ended.

1633 Jamestown's monopoly of shipping breached.

1634 James City County established.

1638 Secretary Kemp built the first brick house in Jamestown. By this year, probably, was opened the first slave market in English America.

1639 On January 11, King Charles I gave approval to the calling of general assemblies in Virginia, thereby setting a precedent for all later royal colonies.

1639-44 The erection of the fourth, or first brick, church in Jamestown.

1640-47 Evolution of the Virginia vernacular story-and-a-half house sheathed with clapboards.

1642 Sir William Berkeley began his first term as Governor in February.

1642-48 Jamestown officials persecuted Virginia puritans.

1644 Opechancanough's *fourth* massacre, April 18. Late in October the great chieftain was assassinated in Jamestown.

1645 Jamestown was allowed its own burgess in the Assembly.

1652 Governor Berkeley surrendered Virginia to the Parliamentary fleet lying before Jamestown on March 12. From April of this year until 1660 the burgesses dominated the colony's government.

1656 The first Quakers arrived in Virginia.

1660 On March 3 the Assembly chose Sir William Berkeley to be Governor again; he assumed office on March 21.

1661 In March the General Assembly was elected and was prorogued successively until 1676. Jamestown was accorded one burgess and James City County two; all vestries were limited to twelve members and made self-perpetuating. William Robinson, Quaker preacher, jailed at Jamestown.

1662 Jamestown ceased to be the sole port of entry for the colony.

1676 Bacon's Rebellion broke out on the Virginia frontier. Berkeley called the first election for the Assembly in fifteen years, May 10. On June 5 Bacon held a night conference with Lawrence in Jamestown. "Bacon's Laws" were passed by the Assembly, June 6-25, during the "rebel's" absence. On June 9 Bacon submitted to the Governor and Assembly, and on June 23 Bacon returned to Jamestown, where he was appointed commander against the Indians. Two days later, the Governor reluctantly signed "Bacon's Laws." At Middle Plantation on July 29, Bacon issued "The Declaration of the People." Berkeley's attack on Bacon at Green Spring failed, September 15; four days later Bacon's forces by a ruse entered Jamestown and burned down the town that night. The Governor returned to the burned-out capital November 9.

1677 One thousand English soldiers landed on February 11, and gave Jamestown its greatest population ever.

1680 Once again, April 25, the Assembly met in the partly rebuilt town. In May, Governor Lord Culpeper began to carry out his instructions for the complete rebuilding of Jamestown.

1682-84 The ministry of the Rev. John Clayton at Jamestown.

1682 Mr. Clayton and others formed a Law Society.

1686 On St. George's Day (April 23) a Cockney Feast was held in town.

1689 Accession of William and Mary was celebrated April 27.

1690 On July 23 the first convention of Anglican clergy was held in the Virginia capital.

1691 Jamestown was back to normal for the first time since 1676. On April 23, "Olympic Games" were held on St. George's Day—the earliest recorded American athletic competition.

1693 The Assembly passed an act on October 10 to open a college at Middle Plantation (later Williamsburg).

1694   The Rev. James Blair, Anglican Commissary, took his seat on the Council on July 18.

1698   Jamestown was again almost totally burned out on October 21.

1699   Governor Francis Nicholson and the Assembly decided on April 27 to transfer the seat of the government of Virginia to Middle Plantation.

Appendix II

A Note on Archaeology
and Restoration at Jamestown

These are the digging years. Recently thousands of Americans have been sharing with their European brethren the fascinating study of popular archaeology, and they have been thrilled to discover that their land too has been the site of "prehistory." The artifacts, which are being uncovered by diligent use of the spade and sieve, can be interpreted—often reconstructed in part—by archaeologists and anthropologists. *Shard* is now a familiar word.

As long ago as 1893, the Society for the Preservation of Virginia Antiquities sponsored "a dig" to uncover the site of the first fort at Jamestown, and in 1934 the National Park Service undertook the subsurface exploration of the entire Jamestown Island. By the time of the 350th anniversary of the landing of the English, 1957, the archaeologists' trenches had revealed the existence of 140 structures of many sorts—dwelling houses, industrial buildings, public edifices, and outhouses. Early Jamestown was a settlement of predominantly wooden buildings; there were only twelve made of brick. One of the most exciting finds was the foundation of a "row house" with three tenements—the prototype of what was to become the standard American urban dwelling house.

Of first importance in any archaeological undertaking is the *interpretation* of what is uncovered and the *reconstruction* of the buildings. "Rarely has a project been as successful as

the reconstruction of the fort at Jamestown," a noted authority tells us. Because the structures of seventeenth-century
Jamestown were erected at various times and no substantial
number fitted a given period, the officials of the National
Park Service very wisely decided not to attempt to reconstruct the settlement as a whole to represent any single period.
The foundations of the other structures located have been
covered over to prevent any deterioration, but all have been
photographed, measured, and otherwise described for future
study.

In the course of the excavations, thousands of artifacts of
many kinds were turned up: metal, clay pottery, beads, coins,
and the like, which have been cleaned, classified, and displayed in what is the largest collection of its kind for any
place of seventeenth-century America. Sparse, though exciting and convincing, evidence was found under the earthen
Confederate fort of the existence long before the arrival of
the English of an Indian settlement—Paspahegh. Everything
uncovered in the digging has been preserved as "raw material"
or "sources" for present and future historians.

Today the glory of Jamestown is the serene and lonely
beauty of the spot, which cannot fail to evoke for the
imaginative visitor a vision of the beginnings of the American people.*

* Three fine books, each in its special way, progressively tell the
story of the archaeological work and its contribution to an understanding of early Jamestown's appearance and the daily life of its inhabitants. One should begin with the very readable and superbly illustrated popular account by Ivor Noël Hume, *Here Lies Virginia* (New
York, 1963). John L. Cotter and J. Paul Hudson summarized for the
general reader in *New Discoveries at Jamestown: Site of the First
English Settlement in America* (Washington: National Park Service,
1957), the extensive, detailed, and authoritative mimeographed report
of the enterprise made by John L. Cotter, *Archaeological Excavations
at Jamestown, Virginia* (Washington, 1958).

Appendix III

For Further Reading

Ten Interesting and Readable Books
about Early Virginia

The following volumes (three contemporary, seven modern) have been selected from among many on the subject, because they seem to meet both the needs and the desires of the lay reader who wishes to know more about 17th-century Virginia.

1. Stephen Vincent Benét, *Western Star* (New York, 1943). This celebrated story-telling poem evokes, as does nothing else, the mood and atmosphere of the London scene and the first English colony.

2. Carl Bridenbaugh, *Vexed and Troubled Englishmen, 1590-1642* (New York and Oxford, 1967); corrected paperback, Galaxy Book (London, New York, and Oxford, 1976). A wide-ranging description of the society that forced Englishmen to emigrate to America.

3. *Travels and Works of Captain John Smith*, edited by Edward Arber and A. G. Bradley (2 vols., Edinburgh, 1910). A new edition, *The Works of Captain John Smith*,

edited by Philip Barbour in three volumes, will be published for the Institute of Early American History and Culture by the University of North Carolina Press, Chapel Hill, in 1980. This is the most detailed record of what went on in the first two decades at Jamestown, vividly narrated by the leading settler who, though small physically, achieved considerable stature by his actions and pen.

4. Philip Barbour, *The Three Worlds of Captain John Smith* (Boston, 1964). The soundest and best informed, as well as one of the most readable, of the biographies of Smith; the author proves that, though he often exaggerated his own role, the Captain was not the liar that 19th-century historians said he was.

5. Ralph Hamor, *A True Discourse of the Present State of Virginia* [London, 1615], facsimile edition (Virginia State Library: Richmond, 1957); also in facsimile (Theatris Orbis Terrarum: Amsterdam, 1971). The Pocahontas-Rolfe story—and much more—related by one who knew them well.

6. Philip Barbour, *Pocahontas and Her World* (Boston, 1970). Not only is this the best and most interesting life of the princess, but it also contains the most up-to-date introduction to the Indians of the Tidewater.

7. Edmund S. Morgan, *American Slavery, American Freedom* (New York, 1975). A fascinating and provocative reassessment of the society of Virginia that has superseded most earlier histories.

8. Robert Beverley, *History and Present State of Virginia* [London, 1722], edited by Louis B. Wright (Chapel

Hill, 1947). A lively account written at the end of the period by a member of the planter-politician-merchant oligarchy, whose knowledge and understanding of the Indians of Virginia was unsurpassed.

9. Abbot E. Smith, *Colonists in Bondage: White Servitude and Convict Labor in America, 1607-1776* (Chapel Hill, 1947). A well-written, definitive account.

10. Ivor Noël Hume, *Here Lies Virginia: An Archaeologist's View of Colonial Life and History* (New York, 1963). Breezy, authoritative, and revealing introduction to "the digging up" of the Old Dominion.

A Special Book

W. P. Cumming, R. A. Skelton, and D. B. Quinn, *The Discovery of America* (New York, 1972). Three eminent scholars discuss the discovery of North America as "seen, experienced, and recorded by Europeans" from the earliest voyages to about 1634. Lavishly illustrated by contemporary charts, maps, engravings, paintings, drawings, letters, and books, this work assists the layman imaginatively to visualize and reconstruct the experiences of the first settlers in Virginia.

Appendix IV *

The Governors of Virginia to 1699

An attempt has been made to give as nearly as possible the dates of actual service of each of the men who acted as colonial governor in Virginia. The date of commission is usually much earlier.

President of the Council in Virginia

Edward-Maria Wingfield, May 14-September 10, 1607.
John Ratcliffe, September 10, 1607-September 10?, 1608.
John Smith, September 10, 1608-September 10?, 1609.
George Percy, September 10?, 1609-May 23, 1610.

The Virginia Company

Thomas West, Third Lord De La Warr, Governor. February 28, 1610-June 7, 1618.

* The list of Governors of Virginia has been reprinted by kind permission of the Jamestown-Yorktown Foundation. The list was originally published in William W. Abbot, *A Virginia Chronology, 1585-1783* (Jamestown 350th Anniversary Booklets, Williamsburg, Va., 1957), 74-75.

Sir Thomas Gates, Lieutenant-Governor. May 23-June 10, 1610.

Thomas West, Lord De La Warr, Governor. June 10, 1610-March 28, 1611.

George Percy, Deputy-Governor. March 28-May 19, 1611.

Sir Thomas Dale, Deputy-Governor. May 19-August 2?, 1611.

Sir Thomas Gates, Lieutenant-Governor. August 2?, 1611-c. March 1, 1614.

Sir Thomas Dale, Deputy-Governor. c. March 1, 1614-April?, 1616.

George Yeardley, Deputy-Governor. April?, 1616-May 15, 1617.

Samuel Argall, *Present* Governor. May 15, 1617-c. April 10, 1619.

Nathaniel Powell, Deputy-Governor. c. April 10-18, 1619.

Sir George Yeardley, Governor. April 18, 1619-November 18, 1621.

Sir Francis Wyatt, Governor. November 18, 1621-c. May 17, 1626.

Royal Province

Sir George Yeardley. May?, 1626-November 13, 1627.

Francis West. November 14, 1627-c. March, 1629.

Doctor John Pott. March 5, 1629-March?, 1630.

Sir John Harvey. March?, 1630-April 28, 1635.

John West. May 7, 1635-January 18, 1637.

Sir John Harvey. January 18, 1637-November?, 1639.

Sir Francis Wyatt. November?, 1639-February, 1642.

Sir William Berkeley. February, 1642-March 12, 1652.

(Richard Kemp, Deputy-Governor. June, 1644-June 7, 1645.)

The Commonwealth

Richard Bennett. April 30, 1652-March 31, 1655.
Edward Digges. March 31, 1655-December, 1656.
Samuel Mathews. December, 1656-January, 1660.
Sir William Berkeley. March, 1660.

Royal Province

Sir William Berkeley. March, 1660-April 27, 1677.
(Francis Moryson, Deputy-Governor. April 30, 1661-November or December, 1662.)
Colonel Herbert Jeffreys, Lieutenant-Governor. April 27, 1677-December 17, 1678.
Thomas Lord Culpeper, Governor. July 20, 1677-August, 1683.
(Sir Henry Chicheley, Deputy-Governor. December 30, 1678-May 10, 1680; August 11, 1680-December 1, 1682.)
(Nicholas Spencer, Deputy-Governor. May 22, 1683-February 21, 1684.)
Francis, Lord Howard, Fifth Baron of Effingham, Governor. February 21, 1684-March 1, 1692.
(Nathaniel Bacon, Sr., Deputy-Governor. June 19-c. September, 1684; July 1,-c. September 1, 1687; February 27?, 1689-June 3, 1690.)
Francis Nicholson, Lieutenant-Governor. June 3, 1690-September 20, 1692.
Sir Edmund Andros, Governor. September 20, 1692-December 9?, 1698.
(Ralph Wormeley, Deputy-Governor. September 25-c. October 6, 1693.)
Francis Nicholson, Governor. December 9, 1698-August 15, 1705.

Index

Agriculture and crops, 34, 40-43.
  See also Corn; Flax; Indians,
  agriculture; Tobacco; Wheat
Ajacán (homeland of Opechan-
  canough), 14, 15
Algonkian tribes, 9, 10, 25, 28
America, Exploration of. *See*
  English invasion of Virginia;
  European exploration of New
  World
Anglicans, 62, 66. *See also* Church
  of England in Virginia
Archer, Gabriel, cited, 19
Argall, Gov. Samuel, 23, 133, 134,
  143; conditions in Jamestown
  when he arrived, 41, 69, 132,
  135; kidnapping of Pocahontas,
  22, 36; cited, 109, 147
Arnold, Anthony, 94
Assembly, General, 77-82, 87-88,
  91, 110
Atkins, Mr., 54

Bacon, Nathaniel, Jr., 90, 91;
  leads frontiersmen against Indians, 91, 98; rebellion becomes
  war for government reform,
  89, 92-103; "The Declaration
of the People," 99; "Nathaniel
  Bacon Esq'r His Manifesto
  Concerning the Present Troubles in Virginia," 94-96
Bacon, Nathaniel the Elder, 124
Bennett, Gov. Richard, 138
Berkeley, Gov. Sir William, 55-
  56, 83-84, 86; housing problems, 137-39, 141; rebels' complaints, 90, 94-95, 99; religious
  regulations and persecutions,
  66, 67, 68; suppresses Bacon's
  Rebellion, 87, 91-103; cited, 55-
  56, 60, 70, 146; *Discourse and
  View of Virginia*, cited, 42
Best, Thomas, cited, 54
Beverages, Alcoholic, 122-23
Beverley, Robert, 87; cited, 127,
  150
Bishop proposed for Virginia,
  72-73
Blacks, 51-53. *See also* Slaves and
  slavery
Blair, Rev. James, 73-74; cited,
  124, 150
Bland, Giles, 93, 94, 102
Brent, Giles, 101
Buck, Rev. Richard, 37, 62, 78

Burgesses, 78, 79, 84-85; for James City, 114, 115; for James City County, 114
Burgesses, House of. See Assembly, General
Byrd, William, 147

Capps, William, 46
Catholics, Roman, 67, 68
Chanco, an Indian, 29
Charles I, of England, 67, 82, 153
Charles II, of England, 87, 89, 107, 117
Charles City, Va., 109
Chew, John, 58
Chickahominies, 23, 24-26
Christianizing Indians, 26-27, 64
Church of England in Virginia, 62, 65-66, 69-70, 72-75. See also Parishes and vestries
Churches of Jamestown, 69, 102, 129, 133, 137, 142
Churchwardens, 112-13
Civil War and Commonwealth, 86, 114
Claiborne, William, 134
Clayton, Rev. John, 48, 70, 123
Cole, William, 67
College of William and Mary, 74
Colonists. See Englishmen; Virginia, Colony of, early settlers
Colonization, Motives for, 4-6, 10. See also Virginia Company of London
Compton, Henry, 83
Corn, 35, 41. See also Englishmen, demand provisions
Coronas, Capt. Pedro de, 14
Corporation of James City, 109-10, 112, 122
Council of State, 81, 82-84, 113
Counties, 111. See also James City County
Court House, 117
Courts, 42, 110, 111, 117, 119;

ecclesiastical courts, 74; general court, 83, 113
Crafts, 41-42, 127, 131
Crashaw, Rev. William, cited, 26
Crops. See Agriculture and crops
Culpeper, John, 92
Culpeper, Thomas, Lord, 120, 141; cited, 55, 70, 142, 148

Dale, Gov. Sir Thomas, 63; attacks on Indians, 50; founding of Henrico, 108, 131-32; negotiates truce of 1614, 22-23; rebuilding of Jamestown, 131; sails for Europe, 24, 38, 77; cited, 121, 131
De la Warr, Lord, 108, 130-31
De Vries, David Pietersen, 146; cited, 42, 46-47, 59, 123
Deaths. See Mortality and disease
Dekker, Thomas, cited, 43
Dominicans, 12, 13, 14
Don Luis. See Velasco, Luis de, an Indian
Drummond, William, 91-92, 94, 97, 98, 101-2, 116, 141; cited, 100
Dutch traders, 51, 146-47. See also De Vries, David Pietersen

Edmundson, William, 68
Education. See College of William and Mary; Henrico, Indian school project
Effingham, Francis Howard, Baron, 87, 123
Elections, 77-78, 91, 115-16
Ellesmere, Lord, cited, 9
English invasion of Virginia, 3-4, 8-9. See also Virginia, Colony of, early settlers; Virginia Company of London, plans for colony
Englishmen: arrogant attitude toward natives, 26-27, 64-65;

attempts to placate Indians, 25-26; demand provisions, 20, 22, 23-24; encroachment on Indian lands, 26, 50, 53; first meeting with the natives, 19-22; war on Indians, 25, 32. *See also* Virginia, Colony of, early settlers
European exploration of New World, 4-7. *See also* Spanish voyages to New World

Families, Importance of, and effect of lack of, 48, 57, 64, 125
Famine and starvation, 45, 49, 130
Fernández, Domingo, 14
Ferries, 148
Fires. *See* Jamestown, fires
Fitzhugh, William, 123
Flax, 43
Fortifications, 4, 69, 71, 128
Franchise. *See* Elections
Frethorne, Richard, cited, 144

Gambling, 123
Gardiner, Samuel Rawson, cited, 62
Garroway, Mr., M. P., cited, 44, 56
Gataker, Thomas, cited, 64
Gates, Gov. Sir Thomas, 45, 129-30, 134; institutes military government, 107-8; outlines duties of ministers, 111-12
Gentry. *See* Merchants and planters
Godwyn, Rev. Morgan, cited, 116
Gonzáles, Vincent, 15
Gouge, Rev. William, cited, 64
Government: of colony under the Company, 26, 76-77, 107-8, 109, 112-13; under the Commonwealth, 86, 114; under the Crown, 80, 113; of Jamestown,

107-9, 117. *See. also* Assembly, General; Corporation of James City; Council of State; Courts; Governors; *Lawes Divine, Morall and Martiall;* Parishes and vestries; Plutocracy
Governors, 79, 81, 87, 107
Granary, 112
Green, Rev. Roger, 72; *Virginia's Cure,* cited, 69

Hakluyt, Richard, 8; *Voyages,* 7-8
Hamor, Ralph, 43, 63; 131; cited, 36, 40, 108
Harrison, William, 141
Hartwell, Henry, 123
Hartwell, Capt. William, 52
Harvey, Gov. John, 58, 69, 136, 137; cited, 82
Headrights, 26, 59, 83
Health. *See* Mortality and disease
Henrico, 108, 132; Indian school project, 26, 65.
House of Burgesses. *See* Assembly, General
Houses and housing, 120, 129, 131-43; "Virginia forme," 138
Hyde, Nicholas, 54

Indians: agriculture, 34-35, 50; attacks on white men, 26, 28-29, 90-91, 130; driven from peninsula, 33, 51; fear and resentment of white men, 17, 22, 27, 34; trade with colonists, 27-28, 90. *See also* Englishmen; "Massacre" of 1622; Opechancanough; Velasco, Luis de, an Indian; *and names of specific tribes*
Inns and taverns. *See* Public houses
Itopatin, 25

Jack of the Feathers, 28-29
Jackson, Goodman, gunsmith, 144
Jail. *See* Prisons
James I, of England, 6, 7
James II, of England, 87, 124
James City County, 111
Jamestown, 118-19, 127-28, 131-41; arrival of first white men, 3-4, 9, 128-29; death and disease in, 44, 48, 49, 130; effects of "Massacre" of 1622, 41-42; fires, 69, 74, 101-2, 117, 120, 129, 141; proposals to abandon site, 116, 117, 129-30, 133, 137; seat of government, 42, 68-69, 83, 110-11; site and port, 46-47, 118, 127-28, 129-31, 143-48; water supply, 128, 129, 131. *See also* Churches of Jamestown; Government, of Jamestown
Jenkins, Sir Leoline, 72
Jesuits, 12, 15, 17

Kecoughtan, Va., 3, 35, 109
Kemp, Matthew, 136; cited, 54

Land, 26, 42-43. *See also* Englishmen, encroachment on Indian lands; Plantations and landed estates
Language and speech, 127
Law club, 123
Lawes Divine, Morall and Martiall, 79, 108, 111, 112, 131
Lawrence, Richard, 90-91, 93, 95-103
Laws, Bacon's, 97, 115, 116
Lee, Col. Richard, cited, 94
Livestock, 41, 52, 71, 131, 149
Ludwell, Philip, 84
Ludwell, Thomas, 86; cited, 73, 96, 140

Luis, Don. *See* Opechancanough; Velasco, Luis de, an Indian

Market, 148
"Massacre" of 1622, 29; aftermath, 31-32, 41-42
Mathew, Thomas, cited, 90, 91, 93, 96
Menefie, George, 52, 58-59, 137
Menéndez de Avilés, Pedro, 11-14, 16
Merchants and planters, 42, 52, 59-60, 80, 82-83
Middle Plantation (Williamsburg), 75, 116, 157
Ministers, 66, 69, 72, 74; puritans, 62-63. *See also* Parishes and vestries
Mortality and disease: Englishmen, 29, 43-49, 55-56, 130; Indians, 49-51. *See also* Jamestown, death and disease
Moryson, Francis, 86; cited, 72, 85, 141
Murray (Moray), Dr. Alexander, 73

Navigation Act of 1651, 147
Necotowance, 43
Nemattanow. *See* Jack of the Feathers
"New Towne," 133-34, 135
Newport, Capt. Christopher, 6, 7, 9, 19, 128
Nicholls, Thomas, cited, 125
Nicholson, Gov. Francis, 73-74, 75, 124; cited, 60
Noell, Martin, cited, 138-39
Northwest Passage, 5, 8
Norwood, Henry, cited, 97

Occaneechees, 91
Oldmixon, John, cited, 121, 123

"Olympic Games," 124
Opechancanough, 10-33; early years in Spain and Mexico under the name of Luis de Velasco, 11-15; return to Ajacán, 15, 16; breaks with Spanish and rejoins Indians and adopts name of Opechancanough, 17; accepted as a brother by Powhatan, 16; death, 33; determines to destroy white invaders, 19, 28, 29; encounters with English, 19-22; tribal diplomacy, 19, 23-24, 25, 27-28, 31. *See also* "Massacre" of 1622; Velasco, Luis de, an Indian
Ordinaries. *See* Public houses
Oré, Father Gerónimo de, cited, 14
Owen, Capt. William, 148
Ozinies, King of, 23

Pace, Richard, 29
Pamunkeys, 19, 22
Parishes and vestries, 111-14, 115-16
Paspaheghs, 3, 10
Percy, George, 108, 109; cited, 35, 45
Philip II, of Spain, 5, 12, 13
Pierce, Jane (Mrs. John Rolfe), 40
Pierce, Mistress (Mrs. William), 126
Pierce, Thomas, 78
Pierce, Capt. William, 108, 110, 126
Piersey, Abraham, 51-52
Plantations and landed estates, 26, 43, 52, 59, 145
Planters. *See* Merchants and planters
Plowden, Sir Edmund, 47
Plutocracy, 80, 82, 88

Pocahontas, 21-22, 36-38, 39-40, 63
Population, 26, 28, 57; Jamestown, 119-21, 127-28, 135
Port of Jamestown. *See* Jamestown, site and port
Pory, John, 78, 110; cited, 118, 126
Potomacs, 31
Pott, Dr. John, 54-55
Powell, Capt. William, 78, 108
Powhatan (brother of Opechancanough), 10, 16, 17, 20, 22, 23, 25
Powhatan empire, 19, 23
Powhatans, 25, 51
Preen, John, 144
Prisons, 67-68, 120, 142
Privateers, 5, 6
Prophecies, portents, and myths, 17, 19, 89
Public houses, 122-23, 136
Purchas, Rev. Samuel, 38; *Purchas His Pilgrimes*, cited, 38
Puritans and puritanism, 36, 61-64, 65-66

Quakers, persecution of, 67-68
Quirós, Father, S. J., 16, 17

Raleigh, Sir Walter, 7
Recreation and social gatherings, 122-24, 131, 151-52
Religion, 5, 10, 61-66; intolerance and persecution of dissenters, 66-68. *See also* Anglicans; Catholics, Roman; Church of England in Virginia; Puritans and puritanism; Quakers
Revel, James, 56-77
Roads, 156
Roanoke, Lost Colony of, 7, 9
Robinson, William, 67

Rogel, Juan, S. J., cited, 13
Rolfe, John, 22, 27, 35, 38; government service, 37-40; marriages, 23, 36-37, 40; puritan bent, 36, 63-64; tobacco grower, 36, 37, 40; *A True Relation of the State of Virginia, Left by Sir Thomas Dale, Knight, in May last, 1616,* 38; cited, 121, 132, 133
Ruling class. *See* Merchants and planters; Plutocracy

San Pedro, Father Pablo de, 13
Sandys, Sir Edwin, 26, 41, 77, 133
Sandys, George, cited, 109
Scarfe, Master John, 108
Scrivener, Matthew, 129
"Seasoning" of immigrants, 46-47
Segura, Father Juan Baptista de, S. J., 15, 16, 17
Self-government, goal of, 26, 77, 79, 85-86, 107
Servants, Indentured, 48, 52, 53-57, 143
Sharpe, Lieut. John, 119, 132
Sherwood, William, 141
Shipping. *See* Trade and shipping
Slaves and slavery, 52, 53, 55
Smith, Capt. John, 20-22, 63, 119, 128-29; *Generall Historie,* 126; cited, 135, 143
Smith, Robert, 86
Smith, Sir Thomas, 133
Social classes, 57-58, 59-60, 124. *See also* Merchants and planters; Plutocracy
Somers, Sir George, 46
Soney, Henry, 113
Spanish voyages to New World, 4-6, 10, 11, 13, 19
Sparkes, Robert, 22
Speech. *See* Language and speech
Spense, William, 78

Sprague, Eleanor, 125
State House, 117, 120, 136, 140, 142, 143
Stevens, Richard, 58

Taverns. *See* Public houses
Taxation, 81-82, 85, 110, 116
Thorpe, Master George, 64, 65; cited, 48
Tobacco, 24, 35, 37, 40-43, 71, 143, 145
Towns, 109. *See also* Henrico; Kecoughtan, Va.
Trade and shipping, 40-43, 143-47
Treaties and agreements with Indians, 22, 23, 24
Twine, John, 78

Velasco, Luis de, an Indian: years spent in Spain and Mexico, 11-15; rebuked by Jesuits, he "goes native," 16-17; massacre of Jesuits, 17; changes his name to Opechancanough, 17
Velasco, Luis de, Viceroy of Mexico, 13
Vestries. *See* Parishes and vestries
Virginia, Colony of, 34, 47, 139; early settlers, 7, 34, 121. *See also* Jamestown; Virginia Company of London, plans for colony
Virginia Company of London, 7, 45, 53, 65; plans for colony, 8-9, 10, 26, 34; revises rules for governing colony, 76-77, 79

Warehouses, 135, 141, 142, 145, 148. *See also* Granary
West, Francis, 108
West, Gov. John, cited, 46
Wheat, 43
Whitaker, Rev. Alexander, 36-37, 62-63, 64; cited, 27, 131

Williamsburg. *See* Middle
Plantation
Wilson, George, 67-68
Women, 48, 53, 125-26
Wyatt, Gov. Sir Francis, 28, 29,
42, 79, 80; cited, 48, 110
Wyatt, George, cited, 29

Yeardley, Gov. Sir George, 24,
28, 51, 58, 59; project "New
Towne," 133-34; promotes tobacco planting, 24, 40-41, 132;
relations with Indians, 24-25, 26;
reorganizes colony in line with
Charter of 1618, 76-78, 109

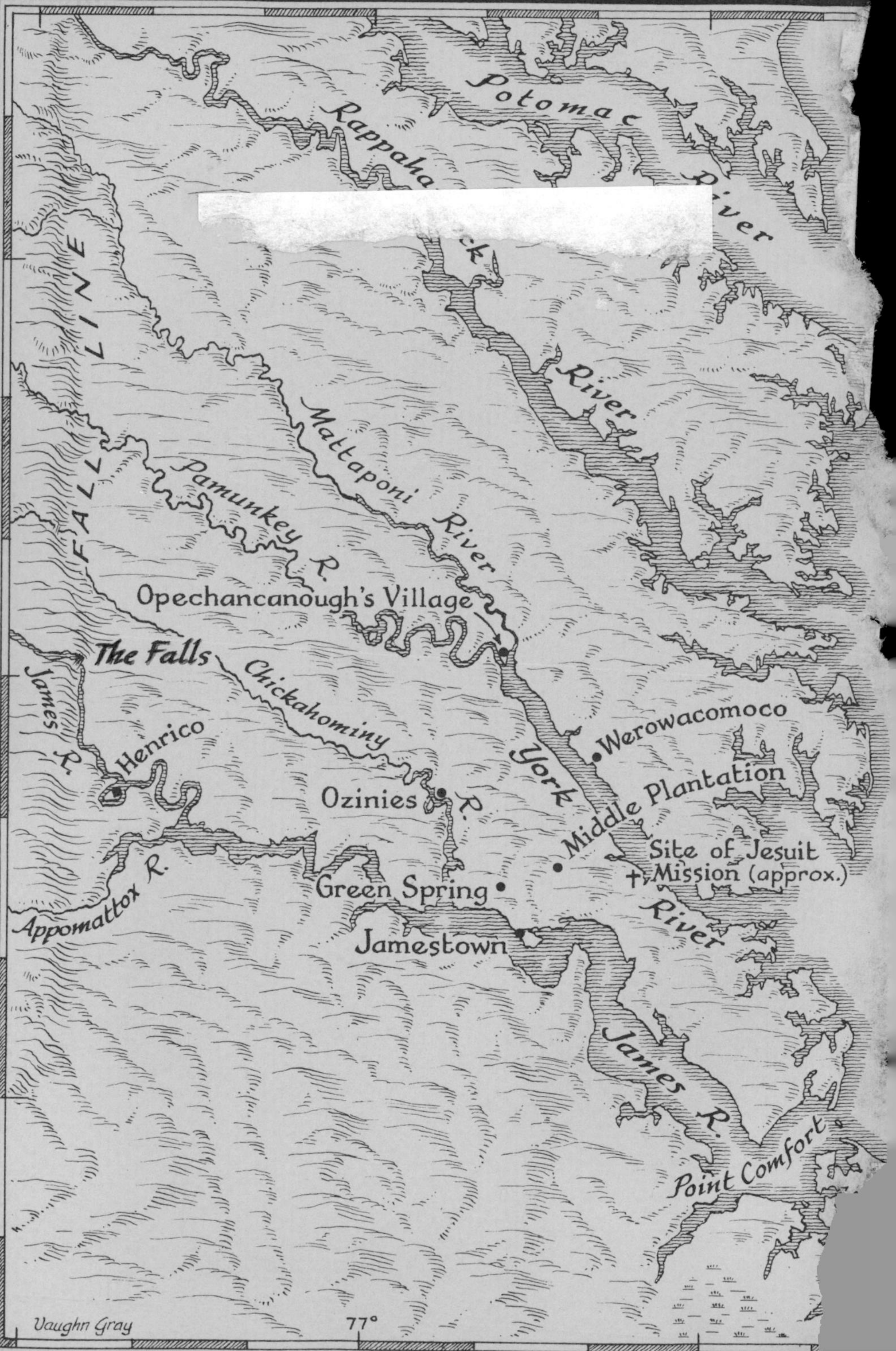
Potomac River
Rappaha
ck
River
FALL LINE
Mattaponi River
Pamunkey R.
Opechancanough's Village
The Falls
Chickahominy R.
James R.
Henrico
Ozinies
York
Werowacomoco
Middle Plantation
Site of Jesuit Mission (approx.)
Green Spring
River
Jamestown
Appomattox R.
James R.
Point Comfort
Vaughn Gray
77°